THE EQUICENTRAL BOOK 2

Healthy Land, Healthy Pasture, Healthy Horses

Jane Myers and Stuart Myers
Equiculture Publishing

Copyright © September 2015

ISBN: 978-0-9941561-8-1

Email: stuart@equiculture.com.au

Disclaimer

The authors and publisher shall have neither liability nor responsibility to any person or entity with respect to any loss or damage caused or alleged to be caused directly or indirectly by the information contained in this book. While the book is as accurate as the authors can make it, there may be errors, omissions and inaccuracies.

About this book

If you watch horses grazing pasture, you would think that they were made for each other. You would in fact be correct; millions of years of evolution have created a symbiotic relationship between equines (and other grazing animals) and grasslands. Our aim as horse owners and as custodians of the land should be to replicate that relationship on our land as closely as possible.

In an ideal world, most horse owners would like to have healthy nutritious pastures on which to graze their horses all year round. Unfortunately, the reality for many horse owners is far from ideal. However, armed with a little knowledge it is usually possible to make a few simple changes in your management system to create an environment which produces healthy, horse friendly pasture, which in turn leads to healthy 'happy' horses.

Correct management of manure, water and vegetation on a horse property is also essential to the well-being of your family, your animals, your property and the wider environment.

This book will help to convince you that good land management is worthwhile on many levels and yields many rewards. You will learn how to manage your land in a way that will save you time and money, keep your horses healthy and content *and* be good for the environment all at the same time. It is one of those rare win-win situations.

Thank you for buying this book and please consider either leaving a review or contacting us with feedback, **stuart@equiculture.com.au**.

About the authors

Jane Myers MSc (Equine Science) is the author of several professional books about horses including the best selling book **Managing Horses on Small Properties** (published by CSIRO).

Jane has lived and breathed horses from a young age and considers herself to be very fortunate in that she has been able to spend her life riding, training and studying these amazing animals.

Stuart Myers (BSc) has a background in human behaviour and has been a horse husband for more years than he cares to remember.

Jane and Stuart are particularly interested in sustainable horsekeeping practices and issues, such as low stress horse management that also delivers environmental benefits. They present workshops to horse owners in Australia, the USA and the UK about sustainable horse and horse property management as part of their business, **Equiculture**.

Their experience is second to none when it comes to this subject as they keep up with recent advances/research and are involved in research themselves. They travel the world as part of their work and they bring this information to you via their books, online resources and website.

See the **Equiculture website www.equiculture.com.au** where you will find lots of great information about sustainable horsekeeping and please join the mailing list while you are there!

Jane and Stuart also have another website that supports their **Horse Rider's Mechanic** series of workbooks. This website is **www.horseridersmechanic.com** why not have a look?

Photo credits

All photos and diagrams by Jane Myers and Stuart Myers unless otherwise accredited. Any errors and omissions please let us know.

Contents

Chapter 1: Introduction

When you watch your horse graze, you are correct in thinking that this is the most natural way for your horse to feed itself. Millions of years of evolution have created this harmonious, symbiotic relationship between grazing animals (your horse) and plants. Plants also benefit from this relationship, just as the horse relies on plants for nutrients those plants rely on the action of grazing to promote growth and reproduction.

However, several factors, including the domestication of horses and the selective breeding and production of certain plant species, have created an imbalance in this once harmonious relationship.

—

Pasture is land used for grazing. Pasture usually consists mainly of grasses with legumes and other forbs (non-grass herbaceous plants).

—

Symbiotic relationship between grazing animals and pasture plants.

On the subject of horses and pasture, there are two real issues of concern for the modern horse owner; the first being how the nutritional changes associated with the new grasses have detrimentally affected the horse, and the second being the challenge that horse owners face to keep horses on relatively small areas of pasture, without causing overgrazing and land degradation.

What we are currently experiencing in the horse world is a growing epidemic. This epidemic has numerous causes and several symptoms, sometimes appearing to conflict with each other, but in reality adding to each other.

The issues include, in no particular order:

- An increase in equine obesity with its related issues.
- Increasing urbanisation and conflict for space.
- The impact of horses on the environment.
- The impact of social media, with peer pressure from its host of inexpert 'experts'.
- Commercialisation, designed to sell a quick fix to the unwary.
- Developers and real estate agents promoting unsuitable properties to horse owners.

Once horse owners appreciate that the causes behind many of these issues are linked, and they understand the relationships between their horses, their pasture and the land on which they are kept, they are able to make informed decisions and make beneficial changes, which are often simple to implement.

Whilst many individual horse owners are oblivious for the need to change traditional practices, there *is* a growing awareness amongst the equine community that there needs to be a change. People are investing time and energy in trying to solve these issues. Horse owners seek information from various sources including the internet, their peers and role models. Some try out different systems to manage their horses and their weight, many of which are based on some form of restrictive feeding/grazing practices; although much of the current research shows that this stresses the horse and can actually have a detrimental effect. These restrictive practices tend to also put additional pressure on the land, leading to increased compaction, mud/dust, weeds and erosion- in other words *unhealthy land*. This not only has a negative effect on the land, but also on the image of the equine community in the eyes of other members of society. This in turn leads to bad feeling and ultimately legislative changes, which, when it comes to legislation on horse ownership, are on the increase. We have to be pro-active and not allow a situation to occur where this happens; we have to become responsible, sustainable and ethical horse owners.

To date, horse ownership has been less regulated than many other walks of life, but that is set to change. If we do not improve the way that we keep and manage our horses and land, then we risk having people who have little understanding about this way of life, regulate it for us. Even if you do not own your own land, you should be looking to the future and how you can make changes for the better that will protect your way of life, improve the welfare of your horse/s, protect the environment and safeguard horsekeeping for future generations, a true win-win situation.

Restrictive feeding practices tend to put additional pressure on the land, leading to increased compaction, mud/dust, weeds and erosion.

Chapter 2: Horses and pasture

The benefits of pasture

Pasture plants are a vital part of our ecosystem and make up a huge proportion of the earth's land surface, 40% of it in fact. Natural pasture is biodiverse and this means that it contains a large variety of species. A biodiverse pasture has many different plants suitable as forage, of which grass is the most prevalent and, probably, the most important. Grass is the most successful of all the plant families – there are over 10,000 species of grass plant worldwide and it provides the bulk of the feed for large grazing herbivores. Pasture has many benefits for horses, the land/environment and ultimately for you – the horse owner.

A biodiverse pasture has many different plants suitable as forage, of which grass is the most prevalent.

Benefits for horses

- Grass and other pasture plants are what horses have evolved to eat; it is their most natural feed source. Horses eat a wide range of pasture species and, although predominantly grazers, they are also browsers and foragers, supplementing their diet with bushes, trees, herbs, berries and succulents.

- These pasture species have evolved over millions of years to have a symbiotic relationship with the animals that graze upon them. Just as the animals cannot survive without the nutrition that biodiverse pasture provides, the grasslands themselves are reliant on the grazing animals for their survival.

- The correct type of pasture is an excellent feed source for most horses. Horses that are working very hard, lactating or growing may need supplementation with concentrate (hard) feed, but this is easily done. Even pasture that is deficient in certain nutrients can be remedied with the addition of supplements (minerals etc.).

Pasture is an excellent feed source for most horses.

- Grazing horses are able to maintain the correct gut fill required to keep gastric ulcers at bay. Horses need and thrive on a very high fibre diet; without it their gut cannot function properly.

- Grazing horses have their head down and are simultaneously draining their airways and breathing fresh air. This is very important as horses have delicate lungs which rely on the lowered head position to keep them clear.

- Pastured horses generally have a better quality of life than their stabled counterparts. Grazing horses are carrying out a natural pattern of behaviour and they do not develop behavioural disorders such as crib biting and weaving. Those that have already developed these behaviours tend to reduce them over time when they spend time at pasture with other horses.

- Pastured horses have better circulation and better hoof quality due to the continual movement associated with grazing because, while grazing, horses are walking slowly and steadily. Slow steady walking, interspersed with short energetic bursts, is ideal exercise for a horse.

- Pastured horses are exposed to sunlight, enabling them to synthesise enough vitamin D. This of course, is reliant on them not wearing rugs that block sunlight.

—

See *The Equicentral System Series Book 1 – Horse Ownership Responsible Sustainable Ethical* for more information about horse grazing behaviour.

—

Benefits for land and the environment

- Pasture plants are a highly efficient 'carbon sink'. They take carbon out of the atmosphere and 'sink' it into the soil. This is very important in this day and age when excess carbon in the air is attributed to global warming. Pasture transfers carbon into the soil even more rapidly than trees. Like trees, pasture also produces oxygen; in fact an area of healthy pasture of approximately 25sqm (275sqft) produces enough oxygen for a family of four.

- Pasture plants collect and hold water, preventing the soil from drying immediately after morning dew and any rainfall. Without these plants, valuable water is wasted as it runs straight off the land.

- Pasture plants also slow the movement of water across the land, allowing it time to be absorbed into the soil and prevent erosion.

Pasture plants collect and hold water.

- As they grow, the roots of the plants allow air and water to penetrate the soil. Plants provide organic matter for soil; as their roots grow and die back in a continuous cycle, organic matter then builds up in the soil.

- Organic matter carries out many functions in soil – not least of all keeping the soil particles apart thus helping to prevent it from becoming compacted.

- Pasture plants cover, cushion and protect soil. Without this protection, soil becomes further compacted under the heavy weight of large grazing animals.

- Pasture plants hold soil together, protecting it from erosion. Without this protection, loose soil, along with manure, ends up in the waterways, causing pollution. This function is very important, as without soil and clean water nothing can survive.

Benefits for horse owners

- Keeping horses at pasture saves money spent on bedding and time spent on stable chores. Horse owners have a less strict timetable; horses do not necessarily have to be exercised every day. Time spent with a horse can be 'quality time'.

- Pasture is a convenient and relatively cheap form of feed. Money spent on pasture renovation saves money spent on feed later. Pasture is up to ten times cheaper than the next cheapest form of feed – bought-in grass hay.

- Properties with good pasture have a higher value, therefore spending money and time on pasture management is a good investment.

Pasture is a convenient and relatively cheap form of feed.

The importance of biodiversity

Pasture can be made up of one species type (a monoculture, which would only occur in a man-made situation), or many species living together (biodiversity). In nature, there is *always* a variety of species in any ecosystem – otherwise it would not be sustainable. In a natural ecosystem, there are many types of plants, animals and insects that live alongside each other and have symbiotic relationships with each other, meaning that they cannot survive without each other. Increasing biodiversity, therefore, is not just about taking care of grazing animals and the plants that they eat, but it is also about providing habitat for numerous beneficial creatures, including certain insects and insectivorous birds.

In nature, there is always a variety of species in any ecosystem – otherwise it would not be sustainable.

The problem with monocultures

Monocultures are prone to disease and pest invasion. When biodiversity is lacking, chemicals such as pesticides and herbicides have to be relied upon more and more, because certain pest insects and plants become dominant in the absence of their natural predators. The problem is that many of these chemicals, as well as causing damage to the environment, are becoming less effective as their overuse has caused resistance to build up in the plants and insects they aim to eradicate. As a result of this, there is now a lot of interest in looking at natural ways of controlling pest insects and plants.

The benefits of biodiversity

- A good, biodiverse pasture provides a wide variety of plants, providing differing nutritional value to horses. Many of these provide different minerals and compounds that have many health benefits too.

- A variety of species will withstand grazing pressure better, as they peak at different times in the season. This staggered growth pattern provides variation to the horse's diet and offers the horse an element of choice as to what plants to graze on and when.

- This well-managed ecosystem also provides habitat for many other animals, some of which naturally predate on pest species.

A good, biodiverse pasture provides a wide variety of plants, providing differing nutritional value to horses.

Increasing biodiversity

A well-managed horse property can do many things to increase biodiversity. By managing the grazing of your horses, for example, you can increase the organic mass on your land and increase the number of both plant *and* animal species.

Many beneficial insects live in healthy grassland and, by using chemicals such as parasitic worming pastes (anthelmintics) responsibly, you can reduce the damage to beneficial insects such as dung beetles (see the section *Dung beetles and other insects*) and earthworms. The same is true with herbicides; responsible usage means less damage to the environment. By planting trees and bushes, you can provide habitat for numerous animals, such as insectivorous birds and insectivorous bats, that will in turn help to control any pest species (see the section *Trees and bushes as habitat for wildlife*).

It will take time to establish a varied pasture, but eventually you should be aiming for a blend of grasses, legumes, medics, sedges and herbs etc. (see the section *What to aim for* for more information). Horses have evolved to eat a great variety of plants on a daily basis, they need variety in their diet and thrive on it – in the wild, horses have access to a huge variety of plants on a daily basis (see the section *Pasture plants in their natural environment*). In the naturally-living situation, horses do not seek a balanced diet every day; they instead seek that balance over the year, as different food sources become available at different times. Biodiverse pasture copies this behaviour as, throughout the seasons, different plants will be available within the grazing area

In the naturally-living situation, horses do not balance their diet every day; they instead seek that balance over the year, as different food sources become available at different times.

Pasture plants in their natural environment

Many pasture plants, in particular pasture grasses, have evolved to coexist with grazing animals, meaning that a symbiotic relationship exists between them and they each rely on each other for survival. Grazing animals obviously rely on plants mainly for food, but the plants rely on the grazing animals for various functions including reproduction, with their seeds being carried and scattered by the effects of grazing.

- In a natural setting, grazing animals move across a landscape; in a migratory fashion for most ruminants and as part of a 'home range' for equines, biting and eating the plants and trampling the rest. The plants are subjected to a short, heavy period of grazing, followed by a period of rest, which stimulates the plants to re-grow and thicken. The animals also leave small indentations on the land with their hooves, into which drops some of their manure, along with the seeds of the plants that they have consumed in the last few days. Some of these seeds are then able to germinate and start life afresh.

Grazing animals move across a landscape; either in a migratory fashion (most ruminants) or as part of a 'home range' (equines).

- In the natural situation, various species of animals graze and browse the plants. Different animal species favour different plant species (with some cross overs), this results in a more even, cleaner graze. Once all of the animals have passed over that particular area of grassland, the plants get chance to rest and recuperate, with the animals only returning periodically to the same areas.

- If they are managed properly, domestic horses and other herbivores such as cattle, sheep, goats, etc. are very beneficial to grasslands – in fact they are essential to it. Without grazing or artificial intervention such as regular mowing – which after all, only attempts to copy grazing, a grassland will eventually become rank and stop growing and may, ultimately, even turn back to forest in some areas. It is all about controlling the grazing pressure so that the plants get the beneficial effects of being grazed without the negative effects of being overgrazed. In turn, the grazing animals (horses and any other animals) get a more varied diet due to the increase in biodiversity (see the section *The importance of biodiversity*), which is fundamental to all grazing animals.

- What happens to pasture plants in the domestic situation is usually quite different. The practice of set-stocking is more common in a domestic situation (see the section *Set-stocking*) and because this in no way replicates how plants have evolved to thrive, they become 'stressed' and develop 'coping mechanisms' as a response to that stress. This can make the plants dangerous for horses to graze (higher in sugar etc.).

The practice of set-stocking is common in the domestic situation.

Pasture plant characteristics

There are various characteristics about pasture plants that horse owners need to understand so that they can make more informed decisions.

Nutritive value

The nutritive value of a pasture plant is a very important consideration when it comes to horses. Horse owners are often told that horses need pasture that is 'highly nutritious', which can be confusing, because this extensively used term is ambiguous; it can mean high levels of macronutrients or energy such as carbohydrates, proteins, fats and/or high levels of micronutrients such as minerals and vitamins.

Horses do not need highly nutritious pasture in terms of energy. In fact, plants can be far too high in energy for many equines. The most important requirement for horses is low energy fibre. So, when applied to horses, we need to be much more specific because horses do not necessarily need feed that is 'highly nutritious, but do need feed that is high in fibre.

The most important requirement for horses is low energy fibre.

—

Plants have both **non-structural carbohydrates (NSC's)** and **structural carbo-hydrates**. NSC is the term used to describe the sugar and starch content of a feed - the carbohydrates the horse can digest without the aid of micro-organisms. Structural carbohydrates are typically found in the cell wall of the plant and are referred to as fibre.

—

Fibre is fermented in the hind gut which is made up of the caecum and large intestine. In horses, the hindgut is home to a large population of micro-organisms which are responsible for digesting fibre. Similar to all mammals, horses are unable to produce the enzymes needed to break down fibre. As the *microorganisms* break the fibre down they produce volatile fatty acids (VFAs) which the horse can utilise as an energy source. In return the horse provides these microorganisms a constant supply of food and a living environment, another example of a symbiotic relationship. On a high fibre diet, the VFAs produced are weaker acids and so the gut is kept within an acceptable range of acidity. Horses need to satisfy most of their energy demands from fibre, which is hard to digest, not easily digestible NSC's found within many grasses.

NSC's are valuable to the horse as a 'fuel' source, especially to those animals working at top end exercise intensity. In contrast, structural carbohydrates are essential to all horses, working or not. It is a matter of balancing the levels of NSC's with the structural carbohydrate content of the diet and the way the horses digestive system works best.

Non Structural carbohydrates can pass through the digestive system fairly rapidly and thus escape digestion in the correct section of the gut. In these situations it is possible for high energy starch from some pasture plants or cereals to reach the hind gut, a place designed mainly for the bacterial breakdown of Non Structural Carbohydrates or fibre. Starch in the hindgut results in the production of very strong acids which can make conditions too toxic for many micro-organisms. This is when problems such as colic and laminitis can occur. Obesity and many related health issues can also be caused by eating a high sugar/starch diet in excess of the horse's energy expenditure.

Learning how to manage your pasture to minimise the NSC content is a very important yet complex issue for horse owners. The NSC levels vary enormously between pasture plant species, but there is also a huge variation *within* any species, depending on various factors.

Some of the factors (besides the species) that affect the NSC content of grass are:

- **The amount of sunshine**. Sunshine allows the grass to create more NSC, therefore shady paddocks are safer. Cloudy days are also safer than sunny days.

- **The temperature**. Cold weather will cause these grasses to store more NSC. A cold night followed by a sunny day is particularly dangerous in terms of raising NSC levels.

- **The time of day**. As the day goes on the plant becomes higher in NSC. Therefore, around mid-afternoon to early evening the levels will be at their highest. During the night (as long as the night is not too cold) these levels will drop as the plant uses its NSC reserves to regrow.

The amount of sunshine – sunshine allows the grass to create more NSC, therefore shady paddocks are safer. Cloudy days are also safer than sunny days.

- **Grazing management**. Stressed grass will be higher in NSC as the plant builds up reserves so that it can recover when the conditions improve. Over grazing causes plants to be stressed.

- **Too much/too little fertiliser**. Different pasture species require different amounts of nutrition; if a particular species of pasture plant is trying to survive in soil of the incorrect fertility, then it too can become stressed.

- **Drought**. This occurs particularly in hotter countries, but most countries experience periods of drought from time to time.

17

- **Too much water**. When water arrives, if it does not soak in or leave the area quickly, it can damage certain plants that are not able to survive in wet conditions.

- **Plant density**. Plants that are sparse will receive more sunlight on each leaf and will be relatively higher in sugar/starch than plants that are densely packed together because each leaf shades out the next leaf.

So, contrary to what has commonly been thought about grass and horses - grass does not *have* to be lush to be high in NSC. In fact, grass that is growing rapidly can be safer because it is utilising its NSC reserves.

The most dangerous times for 'at risk' horses however are when pasture is high in NSC per mouthful (for example - cold nights followed by sunny days) *or* per acre/hectare (when grass is growing rapidly due to climate, rainfall, irrigation etc.). Therefore, it is even more important that the intake of horses is managed when this is the case, along with simultaneously managing the pasture to keep NSC levels low. Remember - horses are often not receiving enough exercise to warrant eating high energy food.

The most dangerous times for 'at risk' horses is when pasture is high in NSC.

'Improved' verses 'unimproved'

As already stated, the nutritive values of individual pasture plant species varies enormously. Some of this variation is man-made and some of this variation is natural, therefore pasture plants can be classed as 'improved' or as 'unimproved. It is an important consideration whether to aim for improved or unimproved plant species in a pasture intended for horses.

Improved plant species.

The term **improved plant species** usually means that the plants in question have been *selectively bred* to be higher in nutritional quality. What this generally means is that the plants ability to produce sugar/starch has been increased and the plants are lower in fibre. These changes make the plants tastier (and more fattening) and these are then able to produce more meat or milk from livestock. Agricultural scientists have been able to yield huge improvements in the palatability and productivity of certain grass species, in particular the ryegrasses.

Some grass species such as Ryegrass has been 'improved' by selective breeding to produce unnaturally high NSC levels.

Unimproved plant species

The term **unimproved** plant species means that the plants in question have *not* been modified. **These pasture plants tend to be native or naturalised plants:**

- **Native plants**. These are plants that evolved in that area, therefore native plants have evolved to suit the local climate and soil conditions. They are usually naturally lower in sugars/starches and higher in fibre. They also generally prefer less fertile soils. This means that they have the added advantage of saving money and resources on fertiliser. There is yet little research to prove that native plants are better for horses, however, anecdotal evidence and knowledge of what constitutes good pasture for horses means that most native plants are a good choice for horses. Depending on your locality, native plants may be harder to obtain and may be more difficult to establish, but the effort should be very worthwhile.

- **Naturalised plants**. These are plants that have been in an area so long that they have become 'naturalised'. These plants generally fit in well with the local growing conditions and, because they are not causing problems, are not earmarked as weeds. Every country with grasslands has naturalised plants, as plants have been moved around the world along with humans and animals. Some naturalised plants have been in place for so long that it is hard to know just when they arrived. These plants are again, usually lower in sugars/starches and higher in fibre and are therefore worth considering for horse pasture.

Until reasonably recently, 'old-fashioned' unimproved pasture plants were thought to have no real economic value. Most countries have various types of unimproved pasture plants and many of these are now receiving renewed interest among horse owners, because they are now regarded as much safer for horses, *and* by farmers.

Farmers are also becoming increasingly interested in native/naturalised grasses because they cope better with adverse weather patterns, in particular drought. They tend to recover more quickly after drought and have a growth period which differs from improved plants therefore they tend to 'peak' at different times in the growing season, resulting in longer grazing seasons. A further advantage is that these plants do not require fertiliser, in fact it can be detrimental to them, and so save a huge expense.

Unimproved pasture plants are well adjusted to a particular environment and provide habitat for native wildlife, *as well* as providing fodder for grazing animals. Whenever possible, native/naturalised grasslands should be preserved or even recreated for the huge environmental benefits that they provide.

Perennials verses annuals

A healthy, biodiverse pasture contains many species of flora, some of which can either be perennial or annual.

Perennials are plants that, once established, will keep growing year after year. As long as these plants are well cared for (e.g. not overgrazed and are allowed periods of rest), they will replicate themselves. The aim should be to establish perennials on your land as soon as possible so that you have permanent pastures that are easier to take care of.

Annuals are plants that grow, set seed and die in one year; they usually need to be re-sown each year, although some new plants may appear from seeds dropped by plants from the previous year. A particular problem with annuals is that, in the period of the year that they die off, soil is left bare and weeds can take this opportunity to become established.

Perennials can take longer to get established than annuals, so annuals can be sown initially to get a pasture started and to cover soil quickly if there is a chance of soil loss due to heavy rain or strong wind. Some annual plants (such as some of the millets) are classed as 'pioneer species' because they are so easy to grow. Their 'job' is to hold the soil together and prepare it for longer lasting, but slower to establish, perennial plants. Perennials can be sown along with annuals and these will be starting to get established as the annuals die off.

Grass growth characteristics

Plants spread themselves around a region in various ways. Some set seed (grow and release seed) and these seeds are carried by the wind, in water, in bird and animal droppings and in the fur of animals etc. However, plants can spread without setting seed, spreading across the ground in various ways. Grasses generally have three different classifications of growth habit. **They are either bunch-type (Caespitose), Stoloniferous or Rhizomatous:**

- **Bunch-type** grasses spread primarily or entirely by the production of tillers. A tiller is a stem produced by grass plants, and refers to all shoots that grow after the initial parent shoot grows from a seed. Tillers originate from the crown area and grow upward from the base giving the plant a clumpy appearance hence the name bunch-type grass. Seeding rates need to be higher when bunch-type grasses are a significant portion of the seed mixture, because you only get one plant from each seed.

- **Stoloniferous** grasses spread by lateral (sideways) stems, called Stolons, that creep over the ground and give rise to new shoots every so often along the length of the Stolon.

- **Rhizomatous** grasses spread below ground stems known as rhizomes. These rhizomes finish in a shoot that emerges some distance from the mother plant. As these new shoots mature they will also produce rhizomes that eventually produce new shoots thus creating a dense sward. If rhizomes are broken into pieces, each piece may be able to give rise to a new plant.

Stoloniferous grasses spread by lateral (sideways) stems, called Stolons, that creep over the ground and give rise to new shoots every so often along the length of the Stolon.

Grasses with the stoloniferous or rhizomatous habit are good at forming a dense covering, however, if they are an undesirable plant (a weed), they can be very invasive. Keep in mind that even grasses can be weeds – it all depends on what is wanted/needed etc. What is classed as a beneficial plant to one group of land managers can be classed as a weed to another.

Grass physiological groups

Most of the grasses (and plants in general) divide into two physiological groups. They are either C3 or C4.

Cool season grass plants thrive in temperate climates and use what is called the C3 photosynthetic pathway. Cool season plants have growth peaks in spring and autumn as a response to cooler conditions at those times of the year. They stop growing in hot weather, when temperatures reach around 30°C, but will also slow down or stop around 5°C or less.

Warm season plants thrive in subtropical and tropical climates and use the C4 photosynthetic pathway. Warm season grasses have a peak growth period during the summer months, providing ample water is present, as a response to warmer conditions. They can keep growing at much higher temperatures than 30°C, but will slow down or stop around 15°C or less.

Cool-season plants continue to grow in lower temperatures than warm-season plants, but warm-season plants continue to grow in higher temperatures than cool-season plants. Warm-season plants will tend to be stressed when the weather is cold and, conversely, cool-season plants are stressed when the weather is hot.

It is not essential to understand details of why or how grasses and other plants are classed as C3 or C4, but horse owners need to understand the differences between them which relate to horses. An important difference is that cool-season (C3) grasses store carbohydrates as fructan and warm-season (C4) grasses generally store carbohydrates as starch.

Both sugars, fructans and starchs, are classed as NSC.

Grazing grasses that are high in NSC can result in horses becoming dangerously overweight. Keep in mind that modern, exotic grasses have been developed to be high in NSC, so that cattle and sheep gain weight quickly, helping with meat/milk production. Horses, on the other hand, are meant to be athletes, and are not usually kept as meat or milk animals.

With C4 (warm-season) grasses, starch forms in organelles (modified cells) in the grass and when these organelles are full, the plant stops producing starch, meaning that the amount of starch in a C4 plant can only reach a finite level. With C3 grasses (cool-season), fructans can be stored in various places all over the plant, and the plant will carry on doing this if the conditions are favourable, in this case cold nights followed by sunny days, allow it to do so. This measure allows these cool-season plants to survive in colder climates.

Current research shows that fructans are implicated in the development of laminitis in horses and cool season. (C3) grasses can, therefore, be especially problematic when trying to manage horses that have had, or are at risk of laminitis or obesity related disorders. These grasses are particularly dangerous when stressed (e.g. cold nights, drought, overgrazing), because they then store more fructans so that they can make a rapid recovery when the conditions improve.

Horse owners need to keep in mind that there is a higher risk of laminitis with these grasses; in particular when day and night temperatures fluctuate between cold nights and warm sunny days.

This does not mean that horses will not become obese or develop laminitis on warm-season (C4) grasses; if a horse has free access to these grasses when they are in abundance, then they will still overeat and the result will be dangerous levels of NSC being ingested. Even if the NSC per mouthful is lower, a horse can still overeat given enough time.

Rhodes grass, an example of a C4 grass.

Another important factor is that when grasses are grown in climates that they are not adapted to – so for example if a cool season grass is grown in a sub-tropical or tropical climate, the grass can be stressed and, will therefore usually store more NSC as a response - the plant stores more NSC so that it can survive in the 'wrong' environment.

Legumes

These play a special role in a pasture; legumes such as Alfalfa, Clover and the Acacia family, being a good source of protein, are often a good food source to complement grass. Legumes can take Nitrogen from the air and 'fix' it in the soil. They do this by forming a symbiotic relationship with a bacterium known as Rhizobia, which draws nitrogen from the air. These special bacteria live in lumps (nodules) on the roots of the plant and, when the plant dies or is cut back (by mowing or grazing), this nitrogen is released into the soil, making it available for other plants to use. Legumes that can be useful in a pasture include clovers, lucerne/alfalfa and medics.

Legumes and grasses do have different nutrient requirements so, when they are both present in a pasture, it can be difficult to have both plant types function well. For example, legumes generally require higher levels of phosphorus, potassium, and lime than grasses do.

Horses will tend to 'spot' graze more in paddocks with legumes. They may either prefer the legumes whilst ignoring the grasses, or vice versa – it all depends on what mix of plants (legumes and grasses) is in the pasture (see the section *Palatability*). Therefore, it then becomes even more important that good grazing management strategies such as rotational grazing etc. are employed (see the section *Grazing management systems*). Legumes that are ignored by horses, and therefore left to grow tall, should be regularly mowed and spread around the paddock, so that the decomposing cut plants put nitrogen and organic matter back into the soil.

Clover is a Legume.

There seems to be some confusion amongst some concerned owners of 'problem horses' (laminitics etc.) about the sugar levels in legumes, with some thinking that legumes should be avoided, and others thinking that they should be employed for the benefits they provide; good calcium levels, amongst other things. Generally speaking, legumes are not usually higher in sugar than the grasses in the same pasture – and the sugars they contain are about 50% pectin which is believed to be a less dangerous form of sugar. They also have low levels of fructans which is desirable, as research indicates that fructans is implicated in laminitis. Like grasses, the actual levels of sugar in legumes vary with the stage of growth and environmental conditions.

In general, because legumes contain more calcium than grasses, they can be very useful when sown with certain tropical grasses. Some tropical grasses have high oxalate levels and cause a calcium imbalance in horses that graze exclusively on these species (see the section **Nutritional problems with pasture**).

Legumes also tend to be higher in protein than grasses and, as horses need a low protein diet, legumes should not make up the bulk of the diet. They are however very useful as a supplement to a diet of grass.

Clover is another Legume, but there are some possible complications associated with clover not caused by the plant itself, but by mould and fungi that can occur on the clover, especially when it is stressed. The toxin, slaframine, is produced by the Rhizoctonia fungus, which grows on clovers and alfalfa during periods of stress (high humidity, drought, and continuous grazing). Individual horses react differently to the toxin; some have more resistance than others.

Drought/water tolerance

Depending on your personal circumstances, you may need to consider whether a plant is tolerant to excessive conditions, such as drought or waterlogging (or even salinity). Countries that have regular drought cycles need grasses and other plants that can cope with these situations. Generally speaking, native grasses again fit the bill because they are tolerant of conditions common to the area in which they have evolved. The same is true for areas that naturally have lots of water; plants that are native to that area will be able to cope with the wet conditions.

Generally speaking, native grasses again fit the bill because they are tolerant of conditions common to the area in which they have evolved.

Climate limitations

The situation of the land determines what kind of pasture can be grown. You need to find out if you live in a temperate, subtropical, tropical or even arid climate, and you also need to know the average rainfall for your area. Rainfall varies from region to region and even within a region. Your local Department of Agriculture office can usually help with recommendations of suitable pasture species for your area. It is also useful to speak with local experts as they will know the area well. However, keep in mind that unless these people have experience with horses, they will not be aware of the problems that horses can have with certain pasture plants. So, use their advice as a guide to which you can then add more horse specific information.

Strength and persistence

This is an important consideration, especially with large grazing animals such as horses. Some plants have more strength and persistence than others. Such plants are good for areas of land that get lots of traffic, for example gateways and laneways. Native grasses are not always as strong and persistent as improved grasses, which may have been selectively bred for these characteristics.

Native grasses are not always as strong and persistent as improved grasses.

Keep in mind, however, that a grass which is strong and persistent may become a problem (a weed) if it invades areas that you do not want it to. Grazing management has a bearing on how strong and persistent the plants need to be. If horses are kept individually in small paddocks (which is a common but not recommended practice), the plants need to be extremely persistent because they are subjected to daily grazing pressure. If horses are rotated around the land as a herd, allowing each area time to rest and recuperate, then less persistent grasses will be able to thrive. One approach is to cultivate one or more pastures containing particularly strong/persistent grass species. This pasture is more able to withstand heavy grazing pressure in times when the conditions are not suitable for the other pastures on the land to be in use.

Palatability

The term 'palatability' refers to what animals prefer to eat. It is a term commonly used by agronomists and farmers to describe a *desirable* characteristic of a particular type of pasture plant, but you must keep in mind that if a plant is highly palatable, then it *may* also be high in sugar and starch. For farmers this is good – it means that cows or sheep will gain weight quickly – but with most horses this can be a problem.

The subject of palatability is very complex because there are many factors at work which affect palatability, such as:

- **The time of day** - plants tend to have higher sugar levels as the day goes on.

- **The time of year** - individual plants species 'peak' at different times of the year.

- **The plant's stage of growth** - plants become more fibrous as they increase height.

- **An individual horse's previous experience** - horses develop much of their individual preferences when they are young and grazing with their dam.

- **How hungry a horse is** - hungry horses aim for 'gut fill' ASAP so, when they are hungry, they will select different plants than when they are not as hungry.

- **What other plants are in the pasture** - if a pasture has an imbalance of 'tasty' and 'not so tasty' plants, the more palatable types of plants will be eaten while others are ignored. The ones that are ignored might actually be perfectly good plants, it is just that there are more favoured ones available at the time.

If horses are left to their own devices in a pasture, depending on how persistent or non-persistent the palatable plants are, a pasture will eventually become overgrazed and tend to only contain species that can cope with being overgrazed. This is one of the reasons why rotational grazing is so important. See the section *Rotational grazing*.

Horses *can* be very selective and are generally regarded as 'fussy eaters'. It is not yet *fully* understood exactly what horses prefer. Sugar is thought to be addictive, and a horse that is used to only eating high sugar grasses may appear fussy when turned out in a pasture containing lower sugar grasses. Imagine your children doing your weekly shop, the basket would probably be full of chocolate, ice cream etc. and contain little or no fruit or vegetables; horses can be equally driven by their sweet tooth if that is what they have learned to like.

Palatability' refers to what animals prefer to eat.

Feral horses have been seen to favour grazing on *low energy* high fibre plants that are plentiful, rather than seeking out plants of higher palatability that are scarcer and would mean having to travel further. This probably indicates that their major driver is to have a high level of gut fill rather than nutrition, relying instead on their digestive system to extract the nutrients. They try to keep their digestive system filled to an optimum level, e.g. not full enough and the stomach acid is not being buffered, nor is there a steady progression through the hindgut, too full and they lose agility which may be required to evade a predator.

If horses are left to their own devices in a pasture it will eventually become overgrazed.

Pasture plant stages of growth

Plants have three stages of growth that you need to be familiar with in order to manage pasture efficiently.

Stage one is called the 'vegetative' stage and, in the case of grasses, this is when a plant is up to about 5cm (2ins) high. This figure does vary as grass plants have different final heights. Generally, if plants are reduced below 5cm (2ins), they do not have enough leaf area to utilise moisture from rain and dew, and to trap available sunlight and convert it to energy for growth.

Therefore, pasture grass plants will struggle to thrive if they are kept at this height or lower. Exceptions are grasses that are 'lawn' type grasses *as well* as pasture grasses, for example couch grass. Lawn grasses have been selected for their ability to survive when very short, whereas most pasture grasses are not able to cope with being kept permanently short.

If a pasture is not rested when it reaches the vegetative stage, the plants may die as they cannot cope with grazing pressure *and* trying to regrow at the same time. Sufficient tissue must remain on plants to allow for production of carbohydrates and meet the growth and respiration demands of the plant.

Short grass plants also have relatively higher sugar/starch (NSC) levels as they store these nutrients waiting for an opportunity to regrow. Most of this NSC is stored at the base of the plant, as this is the growth point, e.g. the point where the plant produces new stem cells and leaves. Therefore, short grass can actually be more dangerous for fat and laminitic horses, because it has a higher concentration of NSC. Horses are able to eat very short grass, right down to the ground, due to having two sets of incisors and a mobile top lip. Therefore, deliberately keeping horses on short grass in a belief that it will restrict their intake does not necessarily work.

If plants are rested at this stage, they grow new leaves and enter the next stage of growth.

Stage two is called the 'elongation' stage as it is the period when plants start to grow. The plant uses some of the stored carbohydrates to grow new cells, creating more stem and leaves. As the stem and leaves grow, it increases the area available to the plant for photosynthesis. Photosynthesis is a process whereby plants use the energy from sunlight, combined with water and Carbon Dioxide, and turns it into chemical energy in the form of sugars and starch, which in turn fuel the plants growth. The plant forms two or three leaves as it starts to grow. However, as the stem grows taller, the number of leaves stays the same but increase in length, therefore the plant has the greatest proportion of leaf material relative to stem material during the latter part of this stage. At this stage, grass plants are generally between 5cm-20cm (2ins-8ins) in height. It is in this period that grass plants are using their sugars to grow and best able to cope with grazing pressure.

Pasture plants, stages of growth.

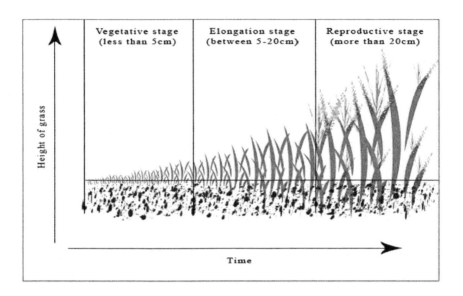

Contrary to what has always been thought with cases of horses that tend to get fat, grass plants are actually safer in many respects when in this stage. It always used to be thought that the best course of action for a problematic (overweight etc.) horse was to put the horse in a paddock of very short grass, with the aim of limiting their intake. However, as we have already seen, horses are more than capable of eating too much, even on very short grass – and remember, this grass is relatively very high in NSC. Longer, rapidly growing grass is using up its sugar and starch reserves to grow and so is less concentrated in these nutrients, with the plant storing most of the sugars at the base. In other words, the plant, particularly further up the stem and in the leaves, has a higher fibre/NSC ratio - the horse has to eat more 'bulk' to take in the same amount of NSC. However, horses that gain weight too easily will still need to have their intake monitored on stage two grasses because they will still overeat, especially if the grass is improved(and therefore high in NSC).

It is very difficult to know the best length for grazing, but current thinking suggests that the optimum time for horses to graze is when the grass has one or two leaves (an average height 15-20 cm). Any less, and the concentration of sugar per mouthful is high and it also stresses the grass; any more than that and the grass is able to lay down extra sugars through photosynthesis. Another optimal stage is later, after the grass has gone through the third stage of growth.

Stage three is called the 'reproductive' stage, once the plant has produced two or three leaves of the optimum length, it starts to maximise its energy production through its leaves via photosynthesis and transitions into the reproductive stage. This stage occurs when grass plants are generally higher than 20cm/8ins. The seed heads develop (set seed) and pollination occurs. Just before this stage happens, the growth point, and therefore the most carbohydrate dense part of the grass plant, moves from the base to further up the plant, where the seed is produced. All new growth in the stem or leaves ceases, and the plant will stay in this condition until an animal or machine (mower) reduces its height.

This plant has reached the reproductive stage of growth and seed heads have formed.

This is the time when hay is traditionally made, as it is the time when the sugar levels within the plant are at their greatest, but it is not the best time to make hay for sedentary horses. Hay for these horses should be made earlier, before the sugars have moved up the stem to produce seed heads, or much later, after many of the seed heads have fallen off (see the section **Conserving pasture** for more information about the best time to cut hay for horses).

Grazing horses on grasses in the reproductive stage can also be dangerous in terms of sugar intake depending on whether the seed heads have just formed or whether they are about to fall off. If they have just formed, the horse will be able to take in lots of sugar as they graze the highest part of the plant. Later, when some of the seeds have fallen off, and the plant has become more fibrous and dormant, the sugar levels drop and it is safer to graze again.

Some horse owners deliberately let the pasture grasses get to this height and, rather than make hay, let their horse/s graze the pasture in the winter. Letting a pasture go to seed is a good way to produce winter feed called 'foggage' or 'standing hay' (see the section *'Foggage' ('standing hay')*).If the grass gets to this height and you want to graze it now rather than later it is a good idea to mow the paddock before allowing the horses to graze it. This puts seeds and organic matter back onto and into the soil and forces the plants back into the elongation stage once more, meaning that they will start to grow again. Periodic mowing of a paddock is a good pasture management strategy for various reasons (see the section *Mowing*).

What to aim for

Good pasture management is about manipulating grazing management systems to ensure that grasses stay in stage two (elongation) for as long as possible. This is done by removing horses when the plant height is reduced to between 5cm (2ins) and 8cm (3ins), just before the plants are about to enter the vegetative stage, and by allowing grazing again when the plants reach between 15cm (6ins) to 20cm (8ins), or when most of the plants have 2 leaves, just before the plants are about to enter the reproductive stage (see the section *Rotational grazing* for more information)

Good pasture management is about aiming to ensure that grasses stay in stage two (elongation) for as long as possible.

Keep in mind that there will be times of the year when pasture does not grow at all because it is dormant (stopped growing). This occurs when it is too cold, too dry or too hot for that particular plant species.

The amount of time that it takes a pasture to recover during a rest period varies with climate, time of year and plant species. There will be times of the year when it is difficult or even impossible to keep all the pasture in stage two of growth at the same time, due to the pasture being dormant or growing rapidly. You will then need to either rest it for longer, (in the case of dormant pasture) mow it, conserve it (make hay/silage) or let it grow, set seed and then mow it later. An alternative to mowing would of course be to use the pasture for grazing later in the season.

If the pasture cannot keep up with the horses, for example during those times of the year when pasture stops growing, or because there are too many horses relative to the amount of available land, then confinement areas such as surfaced yards with shelters must be used for part of each day in order to restrict the amount of pressure that horses inflict on the pasture.

Pasture management is about managing the natural booms and slumps in pasture, alongside managing horses at risk of getting too much or not enough grass.

If the pasture is growing rapidly. You will then need to mow it, conserve it (make hay/silage) or let it grow, set seed and then mow it later. An alternative to mowing would of course be to use the pasture for grazing later in the season.

Misinformation about horses and pasture

When horse owners seek information about pasture for horses there is much confusion about what horses actually need. **This confusion has come about for various reasons:**

- Much of the information in books etc. about pasture and horses still harks back to a time when horses worked for a living *and* plants tended to be lower in energy than they are today. So, for example, it is still common to read that the best pasture for horses is a ryegrass and clover mix. One or two hundred years ago, a working horse certainly needed high energy feed. In fact, there was not enough hours in a day for a horse to work *and* get what they needed in terms of energy replacement from pasture (even if it was relatively high energy) so these horses were stabled, fed ryegrass/clover hay *and* fed additional supplements such as oats.

Much of the information in books etc. about pasture and horses still harks back to a time when horses worked for a living.

- Then came developments in technology and an interest in selective breeding of grasses that met the needs of the cattle industries for beef and milk production. Therefore, modern, selectively bred grasses (and hays made from them) can have extremely high concentrations of sugars such as fructans and starches (NSC's); up to 35% of dry matter in certain conditions. The characteristics that grass breeders select for are higher levels of water soluble (hydrolysable) carbohydrates which include fructan. Fructan in particular is implicated in the equine condition called laminitis.

- The increase in sugar and starch levels in modern grasses has been coupled with a general decline in the work load of horses and an increase in their life span. Horses now live much longer for many reasons such as a lower workload, better parasite control and health care advances etc. These changes have led to huge problems for modern horses. The old information is still out there; confusing horse owners who think that their horses need high-energy feed, yet their horse may not be working very hard at all.

- Most pasture seed suppliers have an agronomy background and are trained in producing highly nutritious pasture for meat/milk production in farm animals. Agricultural science has developed pasture plants that are high protein/high sugar/starch but low fibre, whereas horses require low protein/low sugar/starch and high fibre pasture. When speaking with agronomists, the assumption will generally be that you require a high level of nutrition from your pasture. Agronomists focus on getting the highest possible 'yield' from a given area of land and would therefore usually class unimproved pasture as in need of improvement (see the section **Nutritive value**). If they do have horse experience, it often tends to be with large scale horse properties such as Thoroughbred and Standardbred (racehorse) studs, or other professional large equine establishments that are producing youngstock. Therefore their training and previous experience usually means that they do not understand the very different needs of 'ordinary' horses, unless they have a special interest in them. There is a world of difference between good pasture for young or breeding, high-level performance horses and good pasture for mature, mostly sedentary horses.

- Even equine nutritionists, depending on the background and training, tend to be either 'old school' or more forward thinking and will advise accordingly. In addition, if they currently work for a commercial feed company, their information *may* be biased in favour of focussing on feeding concentrate bagged feed and less on the importance of pasture.

- Commercial horse feed companies – while being a source of some very good information about feeding horses - have a vested interest in leading horse owners to think that their horse needs this or that concentrate bagged feed. This information may be relevant for the few horses that are high level competition or breeding horses, but most are not.

Commercial horse feed companies – while being a source of some very good information about feeding horses - have a vested interest in leading horse owners to think that their horse needs this or that concentrate bagged feed.

- Research about pasture plants is almost exclusively funded by the farming industries, while research about feeding horses is predominantly funded by the performance end of the horse industry, and is carried out by feed and supplement companies. It is understandable that these companies do not tend to carry out any research that indicates that, for many horses, low-energy grasses and hay are the best option, because this is not what they sell. Although many feed companies are now moving towards higher *fibre* bagged feeds, in many cases, all that a horse requires is more low energy hay and/or grass, not a bagged feed.

- Modern horses tend to have a sedentary lifestyle, not moving much at all, they also tend to have little in the way of energy demands due to reproduction, keeping warm or avoiding predators (such as their naturally living relatives do). So modern grasses that have been developed to fatten cattle and to help dairy cows to have high milk yields are far too high in energy for modern horses. These high energy plants can actually be quite dangerous in terms of obesity and its related disorders (laminitis etc.). Also, unlike cattle, horses are expected to live to old age, with many horses now living to their thirties and more.

So when it comes to deciding what grasses are needed, horse owners are in many ways 'on their own'. Horse owners find it frustrating trying to find information about grasses for horses, because in many cases that information either does not exist or is very difficult to find.

In typical human fashion, there has in recent times been a knee-jerk reaction to the phenomenon of horses becoming fat on improved pasture. Many people are incorrectly concluding that, if horses are getting fat on grass, then the answer is to provide an environment where they have access to little or no grass. This can not only cause enormous physical and mental harm to the horse; remember, horses need fibre, lots of fibre, but just as importantly damage to the pasture and the land - again, we will return to this issue later in the book.

Modern horses tend to have a sedentary lifestyle, not moving much at all.

What horses actually need

A mature horse requires approximately 1.5% of its bodyweight per day in dry matter (dry matter = food with the water content taken out), the bulk of this being fibre. Pasture varies in its DM content, depending on such factors as the stage of growth; new grass is lower in DM as it contains more water, older grass is drier and higher in DM. As a general rule, a 500kg (1,100 lbs) horse needs 7.5kg (16.5 lbs) (DM) per day. A horse will usually eat more if it is available, as much as 2%-3% of its bodyweight if given the opportunity. Most horses will take the opportunity to store body fat when feed is available. This is typical behaviour for a mammal.

—

See *The Equicentral System Series Book 1 – Horse Ownership Responsible Sustainable Ethical* for more information about horse grazing behaviour.

—

It's important to remember that fibre is the most important element of the horse's diet. Horses have evolved to use fibre as fuel, providing all the energy horses need for everyday maintenance metabolism--ordinary stuff like breathing, walking, grazing, and sleeping.

It's important to remember that fibre is the most important element of the horse's diet.

Without adequate fibre, the horse's digestive system doesn't function properly; it loses the ability to move food efficiently through the gut. The horse's digestive system has evolved to digest fibrous plant material efficiently, but relies on a steady throughput.

Good pasture will meet the nutritional needs of mature horses at rest and in light work. At certain times of the year, (spring in some areas, the wet season in others) when pasture is growing quickly and has a higher nutritional content, pasture will also feed pregnant mares and young growing horses over one year old. However, lactating mares and weanlings may need supplementary feeding at certain times of the year, such as in the winter or during a dry season.

Pasture yield varies with conditions, ranging from between 80kg (176lbs) of dry matter per hectare (2.5 acres) per day in a high growth period, down to 10kg (22lbs) or even less of dry matter per hectare per day in winter/dry periods. In a drought situation that figure can reduce to zero.

When pasture is growing quickly and has a higher nutritional content, pasture will also feed pregnant mares and young growing horses over one year old.

The actual nutrients in a pasture are also dependent on many factors such as soil type and condition, how often the land is fertilized and the climate etc., but remember, it is as a source of fibre that it is most important. To find out what nutrients are in the soil, you can take soil samples and send them to a soil laboratory to have them analysed. This will tell you what is and is not present in the soil. However, soil tests do not tell you what nutrients the plants themselves contain; to find this out, you can have plant tissue tests carried out. Some laboratories do both types of tests. NSC levels in plants are difficult to analyse because the level varies throughout the day and night. If the pasture is deficient in certain nutrients, then these can either be fed to horses as supplements (e.g. mineral supplements etc.) and/or the soil can be improved over time, but horses will need to be supplemented meanwhile.

It is not always necessary to have pasture that provides all the nutritional needs of a horse; supplementation with minerals etc. is not very expensive in most cases. Often, the most important characteristics of pasture for horses is that it is palatable enough to eat, but not necessarily too palatable as this can mean that it is too high in sugar/starch, and that it is hard wearing etc. See the section *Creating suitable pasture*.

The importance of healthy pasture plants

Healthy pasture plants are desirable for many reasons, but perhaps the most important of these are that they promote a healthy ecosystem and feed your animals. Healthy pasture grows well, provides good ground cover and provides habitat for many species of flora and fauna. The opposite of healthy pasture – 'stressed' pasture - occurs when we ignore important facts about pasture plants.

Many organisms have strategies or defences aimed at dealing with stress and pasture plants are no different. **Some of the things that occur to plants when they are stressed are:**

- They tend to store higher levels of carbohydrate. This is so that they can make as rapid a recovery as possible once the stress is lifted e.g. once it rains again.

- They are more susceptible to parasites or disease, and many of these can be harmful to the grazing animal e.g. fungal growth on certain clovers, as mentioned earlier.

- Some plants give off microtoxins, particularly when being overgrazed, in an attempt to make themselves less palatable and therefore, any animal grazing these plants ingests the microtoxin.

Nutritional problems with pasture

Unfortunately, too much sugar and starch are not the only nutritional problems associated with certain pasture plants. The problems are various and you need to find out what the potential problems are with the pasture plants (or soils) that exist in your locality, or with any new plants that you are planning to sow. You need to then learn as much as possible about those conditions, so that you will be able to recognise and manage any problems and avoid sowing certain problem plants.

Some of these conditions only affect horses, but some affect a wide range of grazing animals. When buying seeds to re-sow a pasture, remember – the seed seller may not be aware of the particular peculiarities of horses (see the section *Misinformation about horses and pasture*). There have been many cases of problem pasture seeds being sold to horse owners by retailers that are inexperienced with horses. You will need to ask around as there are some very knowledgeable people out there – it is just a case of finding them.

Some of the common pasture related problems include:

- Mineral deficiencies occur in pasture plants in some regions (sometimes but not always due to soil deficiencies) and these are usually well understood in the areas that they occur among the local horse owners. These deficiencies can often be corrected with careful supplementation.

- **Tropical pastures and 'Big Head' (Nutritional Secondary Hyperparathy-roidism)** - a condition caused by some tropical grasses with high oxylates. Cases of 'Big Head' have occurred on pastures of tropical species such as Buffel Grass, Green Panic, Setaria, Kikuyu, Guinea Grass, Para Grass, Pangola Grass and Signal Grass. Purple Pigeon Grass is also hazardous. The hazard is greatest when these grasses provide all, or almost all, of the feed available. The oxylates in the grass prevent the absorption of Calcium. This causes weakened bones and bony growths, particularly around the head, but can be prevented with Calcium supplementation, or by the feeding of other forage with low oxylates, including Lucerne (Alfalfa).

- **Ryegrass Toxicity/'Ryegrass Staggers'** - caused by a fungus called endophytes occurring on Ryegrass, although some ryegrasses have been bred to be 'endophyte free'. The toxins in the fungi cause a reaction in the horse's brain causing muscle dysfunction.

- **Equine Fescue Oedema** - another endophyte related disorder, this time affecting winter-active fescues, rather than ryegrasses.

- **Ergot of Paspalum** - another toxic fungal issue; the toxins in this black sticky fungus can cause 'staggers'.

- **Australian Stringhalt** - thought to be caused by a certain weed in certain conditions e.g. during or after a period of drought.

- **Equine Grass Sickness** - mainly occurs in GB and other European countries. It causes damage to the nervous system and is a very serious condition.

- **Nitrate poisoning** - not common, but can be fatal and is caused by ingestion of too much nitrite from pasture or weeds.

—

As these conditions are often common to specific areas we have an **Equiculture website page** that has more information about these conditions and more, with links to further information. See **www.equiculture.com.au/horses-and-toxic plants.html**. If you have any concerns regarding your horse on these or any other issues do not hesitate to consult your vet.

—

See also the **_Poisonous trees and plants_** section of this book.

Paspalum - the toxic fungal issue; the toxins in this black sticky fungus can cause 'staggers'.

Creating suitable pasture

This section is about improving pasture *for horses*. In some cases that will be about increasing the *number* of plants in a pasture, particularly in the case of degraded 'poor' land. Increasing the number of plants in a pasture is achieved through better pasture/grazing management (see the section **Grazing management**). In other cases it will be about changing the pasture plant species over to plants that are more suitable for horses, such as to plants that are lower in energy.

When faced with pasture that appears to be less than desirable, it is common for horse owners, especially if they have just acquired the land, to rush into major renovation projects. Certainly, in terms of halting and reversing land degradation issues such as bare soil, erosion etc. you should act quickly. However, unless you are already experienced at land/pasture management, make sure you spend time learning as much as possible before you make any major changes. In many cases, a few simple procedures will give huge rewards.

First of all you need to take stock and make plans for the future. If your pasture has already reached its potential, then all you have to worry about is maintaining this state. However, on most horse properties, improvements can be made. The information in this section will help you to decide what you need; it gives you a broad outline of knowledge so that you can add local knowledge to get the best results.

In some cases that will be about increasing the number of plants in a pasture, particularly in the case of degraded 'poor' land -before (A)...and after (B).

Before you start

First... find out what you already have (stock take)

If you do not know what pasture plants are currently growing on your land, you need to get them identified as soon as possible. You may find that you actually have desirable grasses already in place. It is very common for people to reseed only to find out later that what they already had was actually better!

Good, biodiverse pasture contains many species of flora so, whilst it is good to have a broad range of plants within the pasture to supplement the horse's diet, as with most things it can be harmful if the balance is skewed and horses are eating a proportionately low ratio of grass to other plants.

Recent studies have shown that many of the plants, commonly thought of as weeds, that horses do like to eat are also high in NSC. That means that they are fine to supplement a mixed diet, as many contain different minerals and chemicals to grasses but, if a horse is living in a very weedy pasture with little or no grass, then their food source will contain a high NSC/fibre ratio, causing the horse to gain weight. Weeds, however you define them, are nature's way of covering bare soil.

Weeds, however you define them, are nature's way of covering bare soil.

Therefore anytime a pasture is bare or degraded, you will have an outcrop of weeds. Some weeds are very beneficial to the soil, their long taproots 'harvesting' minerals etc. from deep underground and also serving to break-up compacted soil. Whilst it is good to have a diverse variety of plant species within your pasture, it should predominantly be made up of grass species, with other plants supplementing this main source of fibre.

There are various ways that you can identify the plants (including weeds) in your pasture:

- **Via the internet**. There are various excellent websites that can help you to identify plants.

———

Many of these are listed on the **Equiculture website Grass/pasture information page**. See **www.equiculture.com.au/grass-pasture-information-links.html**

———

- **A local land/soil conservation group.** Most areas have government funded or voluntary land/soil conservation groups that are a wealth of information about local vegetation and soil conditions. In many cases, representatives will come out to your land for free and help you with plant identification. These groups are usually particularly interested in increasing local and native/naturalised species of vegetation in an area and are usually very pleased to help if you are interested in creating or improving biodiversity.

—

The national bodies of many of these organisations are also listed on the **Equiculture website Grass/pasture information page.** See **www.equiculture.com.au/grass-pasture-information-links.html**

—

Find out about what soil type you are dealing with and what condition it is currently in. Again, a local land/soil conservation group should be able to help. See also the section *Understanding soil.*

Understand your climate

You need to understand the climate that you live in. If you live in a **sub-tropical or tropical** climate, then a concern for you is going to be avoiding pasture grasses that are high in oxalates (see the section **Nutritional problems with pasture**), or managing your land and your horses so that the effects are minimised.

*If you live in a **sub-tropical or tropical** climate, then a concern for you is going to be avoiding pasture grasses that are high in oxalates, such as Seteria.*

In terms of the risk of laminitis and other obesity related disorders, tropical grasses may not be as high in sugar/starch per mouthful, but due to their ability to increase biomass rapidly in 'good' growing conditions, they can be high in sugar/starch per acre/hectare etc. In other words, each plant may be relatively lower in sugar/starch, but the sheer number of plants that a horse has available to them will still mean that the horse is in danger of obesity and all the risks that go with it. So, good grazing management may mean limiting grazing to 'at risk' horses.

If you live in a **temperate climate**, you need to avoid sowing grasses that contain toxins including endophytes (see the section *Nutritional problems with pasture*) and you need to either avoid certain improved pasture plants, or manage them well because of their high capacity for sugar/starch (NSC) accumulation.

Other things to think about

Think about how much land you have access to and how many horses you have now, and are likely to have in the future. See the section *Stocking rates*.

Decide what kind of 'lifestyle' you have, or are planning to create for your horses. Will they be kept in herds or separated? Separated horses (apart from the welfare issues) create land management problems (see the section *Horses and land degradation*). You will need much stronger, more persistent types of pasture plants in some situations.

—

See *The Equicentral System Series Book 1 – Horse Ownership Responsible Sustainable Ethical* for more information about horse 'lifestyle' and how different ways of keeping horses affects land management.

—

Planning your pasture means that you can make informed decisions. Planning will involve taking into account that, in order to have long term productivity, you may have to 'hold your horses' now and spend time and money on renovating the pasture and on supplementary feed. In time, when the land is working more efficiently, you, your horses and the wider environment will benefit hugely.

Plan to do *as little* as possible

- There are different levels of pasture improvement ranging from low input to high input and it is recommend that you start with low input as this is often all that is needed.

- As already mentioned, pasture can often be *hugely* improved by simply improving current management practices. This would involve utilising grazing management systems (see the section *Grazing management systems*) and mowing and harrowing. In addition, soil tests can be carried out and fertiliser added if necessary, along with weed control and bare soil management (mulching and using swales, see the section *Turning land degradation around*). However, you must understand the plants that you have ('good' and 'bad'), because what works to encourage/discourage one type of plant can have the opposite effect on another.

- In many cases, the seeds of the pasture plants you want are *already* present in the soil. They may just need the balance tipping in their favour in order for them to get going. For example, a pasture that has a dense covering of weeds *may* just need *repeated* mowing in order to knock them back and give the grass plants chance to grow (see the section **Control of weeds**).

- Get *local* expert advice. This could be from the local agricultural department, the local land/soil conservation group, neighbours (depending on their level of expertise) or a local rural store; the type of place that sells farm equipment, seeds etc., however these vary enormously in terms of how knowledgeable the staff are.

More extensive renovation

If you are planning extensive renovation of your pasture, it is a good idea to plan to do just one or two areas of your property each year. By doing this, you can gain experience gradually, but also have a place to keep horses while new pastures become established. Renovating all of a given area of land all at once can result in disappointment and wasted money if things do not go to plan. For example, if there happens to be a drought during the year that you implement pasture improvements, there will be poor or non-existent results in the area that you choose. If you have prepared all the land, a drought will be disastrous. All of your soil will be exposed to the air for the duration of the drought, whereas if you have only prepared one part of the land, the rest of the land will still have some vegetation cover through the drought. You need to plan how you will manage the rest of the land because newly renovated pasture will not be able to be grazed for some time, which will put more pressure on the remaining pasture. You will need to confine your horses to part of the land while you carry out renovations; this can either be in surfaced holding yards or a paddock (or part of a paddock) that will be renovated later (a 'sacrifice area') (see the section **'Sacrifice areas'**).

Things to keep in mind are that:

- New pasture needs the correct amount of rainfall to become established; therefore the time of year is critical as there must be enough water both in the soil and from 'follow up' rainfall.

- Weed control; again you will need local expert advice unless you are already experienced.

48

- Have soil tests done to help you determine which area you will renovate first. It is usually better to pick an area that has the most chance of success to begin with, until you gain experience.

- Find out if and when contractors are available; you will need to use a contractor to spread the minerals and fertiliser if you do not have the equipment - a good rural store should be able to help you find someone to do this. Then you can fertilise if necessary (as per soil tests).

New pasture needs the correct amount of rainfall to become established; therefore the time of year is critical as there must be enough water both in the soil and from 'follow up' rainfall.

- Select your species and source quality-certified seed. Remember that the easily available information about which pasture plants to sow for horses may be incorrect (see the section **Misinformation about horses and pasture**).

- Correct soil preparation is very important for successful pasture improvement. Young plants, water and nutrients cannot get through hard compacted soil. However, aim for the least soil disturbance possible.

- Cultivate just before you are ready to seed so that loose soil is not left exposed for a long time. The best time is just before the rains are due so that you can seed the paddocks as soon as it starts to rain. The seeds can then benefit from moisture in the soil and any follow up rain.

- You may only need a small amount of cultivation, e.g. scraping the soil with heavy harrows in order to break the surface of the soil. Seeds can then be broadcast (spread across the surface) by hand or by seed spreader and gently covered with soil by a rake or pasture harrow.

- A higher level of renovation would be to 'sod seed' the pasture, whereby seed is drilled into the soil along with a fertiliser. Most horse owners would need the services of a contractor at this stage. With this method, the old pasture can be left undisturbed if desired, however any established pasture weeds need to be controlled to give the new seeds a good chance. If you wish to retain the established pasture, it can be grazed hard before seeding and for two weeks afterwards to give the new plants a chance to grow. This can be carried out by sheep if they are available, as a large number of sheep will do a better job than a smaller number of say cattle (more mouths – less weight). Repeatedly mowing the pasture is another strategy – it does a similar job to sheep.

- The last resort should be cultivation and reseeding. This is the most expensive solution and can fail if it is not carefully planned and carried out. You may need to remove all of the old pasture by using chemicals. Only consider cultivation if a paddock is currently virtually useless, as it will be out of action for several months to a year after reseeding. If animals are allowed to graze too soon after reseeding, they will pull out the new plants by the roots and compact the loosened soil. Deep cultivation releases air from the soil, so the soil is much more prone to compaction until it builds up organic matter. Weeds must be kept on top of as they will initially be able to establish themselves very easily in the newly disturbed soil. Therefore, consider cultivation as a last resort. If you decide to go ahead, you will need an experienced contractor.

- Any cultivation should only be carried out along contour lines (from one point to another on the same level). If land is cultivated up and down slopes then rain will use these channels as gullies and erosion will result in the newly disturbed soil. Correctly done, cultivation breaks up the flow of water downhill and directs it into the soil.

- Traditional ploughing is not recommended for fragile soils, because a traditional plough turns soil over, rather than leaving the layers in the same order. Deep cultivation, such as traditional ploughing, shatters organic matter, dries out soil, destroys bacteria and fungi that is necessary for plant growth and exposes soil to the air. This 'burns' off organic matter in the soil. Only deep top soils should be ploughed (if at all), because ploughing shallow soils takes the thin layer of topsoil underground. Ploughing can also expose seeds of undesirable plants that have laid dormant.

50

- In paddocks that are very low in nitrogen (as determined by a soil test), consider sowing Legumes either as a crop in the year before the new pasture is sown, or as part of a pasture mix (e.g. Legumes can be added to a pasture sowing mix at the rate of about 0.5kg to 1kg (1.1lb – 2.2lb) per hectare) (1 hectare is 2.5 acres). Sowing a legume such as Lucerne, has many benefits. By repeatedly mowing (or grazing) a plant, the roots die and regenerate releasing nitrogen into the soil. At the same time, the dead roots create organic matter in the soil and allow air and water into the soil.

- Some grasses, for example Australian native grasses, have adapted to soils that are low in nitrogen, so this step may not be necessary if you plan to establish native pastures. However, some grasses can cope with higher levels of nitrogen. Speak with a native seed supplier with an understanding of your particular region before increasing nitrogen levels on your land.

- New pasture seeds can be introduced to the prepared soil when there is a good chance of rain, and more importantly, follow up rain. In small paddocks, one or two acres, it is possible to spread pasture seeds by hand.

New pasture seeds can be introduced to the prepared soil when there is a good chance of rain.

Hold your horses

Once the new plants start to grow, you must 'hold your horses' until it becomes well-established. A test of whether a plant is ready to be grazed is that you should be able to pull and twist the plant and the leaves should break off in your hand. If the plant instead lifts out of the soil, roots included, the pasture is not yet established enough for grazing and needs to be mowed instead. A short, sharp period of grazing with a relatively high number of sheep, if available, is an alternative to mowing.

Putting large grazing animals on new pasture too early will give them little benefit and will prevent the pasture from establishing properly.

Once new pasture is established, you can allow *short supervised* bouts of grazing if the ground is well covered and the plants are not being pulled out of the ground when grazed. It is a good idea to follow up with mowing to 'tidy up' and leave all of the plants at the same height to regrow evenly.

Once new pasture is established, you can allow short supervised bouts of grazing if the ground is well covered and the plants are not being pulled out of the ground when grazed.

It is usually recommended that a newly sown pasture be left a year without grazing pressure, but this is not always necessary if the plants are deep rooted and well established. A certain amount of supervised grazing - just a few hours at a time - and regular mowing (with the blades set high) helps the new plants to thicken and spread. You will have to monitor the situation carefully before you allow horses to graze the new pasture for longer periods.

What to aim for

- In management terms it is easier and more cost effective to have paddocks containing low energy plants that *any* horse can eat safely and then supplement any horses that need it as and when, rather than having paddocks full of plants that are too high in energy. A typical horse property may contain horses of varying size, age, workload and purpose unlike a farmer who would have many animals of the same type. On any given horse property, the usual situation is that only some, if any, of the horses are 'in work' while the others are resting, recovering from an injury, in retirement or the owner does not actually have a plan for them yet. Not to mention those times when an owner cannot ride due their own other commitments etc. Therefore, if the paddocks contain plants that are very high in energy, it can get to the stage where it is not safe to turn horses out - which is inconvenient to say the least and can lead to other problems such as behavioural issues with horses. It is much easier and better for the horses to add feed if and when required than to deprive some or all of your horses because the pasture is too rich.

A typical horse property may contain horses of varying size, age, workload and purpose.

- Different pasture species have many values in horse paddocks. The species you have in your paddocks (or want in your paddocks) varies due to the local climate and soil conditions. There are many available grasses and it is impossible to list them all here, however it is relatively easy to find out what you can grow in your paddocks; speak to a local seed supplier or a local land/soil conservation group, but be aware that they may not understand the special needs of horses. Remember there is a huge range even within a species of how much NSC that plant can contain and it has much to do with the conditions that grass is grown in; always seek local advice, but look for low NSC (average 10% or less if possible), high-fibre grasses.

- Look to encourage 'old fashioned' native/naturalised grasses on your pasture. These species are generally high-fibre and low NSC and have evolved to survive in the local climatic conditions, able to survive drought, floods or extreme temperatures. In many areas, these types of pasture plant are increasingly seen as the way forward for healthy horses.

—

Land conservation groups are a potential source of information about pasture plants. They will generally be in favour of conserving or recreating native or native/naturalised pasture (see the section *Nutritive value*) and this type of pasture is generally what forward thinking horse owners want. If there is a land/soil conservation group (or similar) in your area, this can be a good place to start. They should be able to identify plants on your land, whether they be native/naturalised or otherwise, and help you to obtain good seeds.

—

- If horses are allowed to eat too much of the high-energy pasture, they can end up with laminitis or other obesity related disorders as a result. Not only is laminitis disastrous for the horse and owner concerned, it also has a high financial cost in vet bills etc. Therefore, as stated earlier, it is cheaper and far more humane to supplement those horses that need it with different feed as and when needed than to risk laminitis and other obesity related disorders. Many major horse welfare agencies now regard obesity in horses as more of a welfare issue than underweight horses. Underweight horses can often make a good recovery; overly obese horses rarely do, or are left with permanent secondary issues such as laminitis etc.

- If you would like to establish native/naturalised pasture on your land, it is more difficult than establishing improved species, but it is still possible. First of all you may already have desirable pasture plants growing. As mentioned above, find out exactly what is already growing on your land. These grasses may become more abundant once you start to employ better pasture management. Due to being less persistent, many of the desirable plants cannot cope with over grazing but if the area was previously grassland, as opposed to forest, there should be desirable seeds already in the soil – waiting for the right growing conditions.

- However remember that the soil may also have been changed over the years to favour the more developed species. Over time and with careful management it may revert back to conditions which will suit native/naturalised grasses. This can also be assisted by application of certain minerals following an independent soil test.

- It is possible to buy native/naturalised/ pasture seeds. They tend to be expensive compared to improved species, but they should be a good investment. In time, more horse people will be looking to buy property that has this kind of pasture and will be avoiding properties that have problematic plants e.g. the improved species.

- It is sometimes possible to collect desirable seeds from grasses that are growing around the outside of paddocks. Again, get local advice about when this should be done for the best results.

However remember that the soil may also have been changed over the years to favour the more developed species.

- General 'rural stores' and horse feed stores sometimes sell bags of 'horse pasture seed mix'. These mixes are often too general and may not even suit local conditions etc. They may actually contain undesirable seed types. This is because what is traditionally regarded as good horse pasture and what we now know in light of recent research are two very different things. Ask before you buy such a mix and, if the store does not appear to have expertise in this subject, buy from somewhere that specialises in pasture seeds rather than in horse or general agriculture store, that just sells pasture seeds as a side-line.

- A good rural store specialises in selling seeds. A local rural store will understand the local soil types and climatic conditions and they can usually also recommend local contractors etc. However, make sure you do your homework about which seeds to buy and keep in mind that often agronomists do not have experience of horses. Horses have very different needs to cattle when it comes to selecting pasture species. In fact, recreational horses are at the opposite end of the scale to cattle in terms of what is needed from pasture.

Horses and land degradation

The effects of land degradation are many and they often have repercussions far beyond the immediate surroundings of the land in question. The easily visible signs of land degradation are bare soil, dust, mud, weed infestations and soil erosion, as well as dead or dying trees. Land degradation is a common sight on land on which horses are kept. The effects are not always so easy to see at first glance, and even if they can be seen, they are not necessarily attributed to land degradation.

The easily visible signs of land degradation are bare soil, dust, mud, weed infestations and soil erosion.

Land degradation has many negative effects including:

- Horse health – dust and mud cause respiratory and skin conditions in horses and the only things able to grow in degraded soil are weeds, some of which may be harmful for horses. Even if they are not directly harmful, they should not be the main source of nutrition.

- Soil loss- top soil is being lost at an alarming rate in many countries around the world. For example the UK has lost 80% of its topsoil since 1850 and is still losing it at an average rate of 1-3cm per year!

- Soil quality – soil that is unhealthy is not able to grow healthy plants. Unhealthy soil is generally compacted, has a mineral imbalance and is lacking in microorganisms.

- Water quality - any soil, manure or other pollutants that enter a watercourse have an impact on the quality of the water further along the water system and ultimately on the seabed. For example, soil washed from the land can cover and restrict the growth of seaweed/sea grasses, causing problems for aquatic animals and encouraging the spread of toxic algae.

- Land value – degraded land is worth less financially.

- Your budget - you will need to spend more money on 'bought in' feed.

- Reduced wildlife habitat – this leads to other problems such as large numbers of pest insects due to a shortage of birds or insectivorous bats to eat them. In turn this means that more pesticides have to be used.

Soil loss- top soil is being lost at an alarming rate in many countries around the world.

Farmers raising livestock cannot afford for their land to become degraded as it is this land that provides the most cost effective feed for their business. It makes absolutely no economic sense for a farmer to not maximise the productive capacity of their land, yet horse owners often do this. Horses are not always kept for economic reasons, but this should not be an excuse for mismanagement.

A horse property that has degraded land leads to criticism of the horse industry which, over time, can lead to increased legislation restricting horse ownership in certain areas. This is yet another reason why it is vitally important that land used for horses is managed in such a way that it does not become degraded and an eyesore. As a landowner/land manager you are actually a custodian of the land and have a responsibility to manage that land to the best of your ability. Horse owners have to see a piece of land not just as somewhere to turn horses out in, but recognise its value to not only themselves, but also the wider community and environment.

Bare/compacted soil and erosion

A typical order of occurrence is overgrazing, combined with too much hoof pressure, (these two factors combined are sometimes referred to as *Grazing pressure*), bare soil or mud/dust, leading to soil compaction and, ultimately, soil erosion. Too much grazing pressure means that pasture plants die out – or are worn out. Usually, the plants that disappear are the ones that you most want to keep. This downward spiral of land degradation can be accelerated further by climatic conditions such as drought or extended wet periods when there is either little new plant growth or the soil is too wet.

The sheer weight of grazing animals compacts soil and, without the cushioning effect of plants, the problem is increased; think of plants as a carpet that absorbs some of the impact of animals. Added to this, once plants have died and their roots have shrivelled, there is less organic matter between the soil particles to keep the soil 'open'. As a general rule of thumb, the more plant growth above ground, the deeper and healthier the root system, which helps to keep the soil particles apart and prevents compaction. Once the root system dies back, soil particles pack together increasingly tightly, preventing air and water from entering the soil. At this stage, even if the animals are removed, the only plants that will be able to grow are very hardy weed species with strong tap roots. Certain weeds are opportunistic plants that thrive in conditions that more favourable plants (such as pasture grasses) cannot, hence the proliferation of weeds on many horse properties.

Due to the fact that water cannot penetrate compacted soil, when it rains, it runs off and takes any loose topsoil and manure with it. If the land is sloped, this further adds to the problem, as water will run over bare land more quickly and cause erosion channels. These areas are also prone to further erosion when the weather dries out and they become dusty. The wind then removes yet more topsoil. So, it can be seen that the situation becomes a downward spiral.

Bare, compacted soil typically occurs in high traffic areas such as yards, laneways, near gates, around feed and water points, and shady areas such as under trees. However, if too much grazing pressure is put on land, it can happen to the whole paddock or indeed the whole property.

A result of compaction, even after heavy rain the water is unable to penetrate the soil.

Horses, if left to their own devices, will tend to create bare areas more readily than other grazing animals for various reasons. They can eat closer to the ground than cows, because they have two sets of incisors which are designed to cut plants at a low level and they weigh more than sheep (cloven feet also compact the soil less than horse's hooves).Horses also tend to hang around gateways or other areas where they can see human activity if they are being fed on supplementary feed.

—

See *The Equicentral System Series Book 1 – Horse Ownership Responsible Sustainable Ethical* for more information about how horses behave in paddocks and how this affects land management.

—

If it is necessary to reduce the grazing pressure on your pasture, and therefore bare soil, compaction etc., the best way to do this is to adopt the attitude that if the horses are not grazing, then they should not be in the paddock. Sleeping and loafing take up about eight hours of a horse's day (approximately four hours sleeping/snoozing, four hours loafing). Loafing includes playing, mutual grooming and just standing around together. By controlling the amount of time horses spend in paddocks, this behaviour can be manipulated so that horses mainly graze when in a paddock and mainly sleep and 'loaf' when in yards (either separate or one large surfaced holding yard). By adopting good grazing management systems (see the section **Grazing management systems**), and in particular **The Equicentral System** (see Appendix: **The Equicentral System**), paddocks will not become compacted in the first place. **The Equicentral System** has the added advantage that the horses bring themselves in to the surfaced holding yard, saving you time.

There are various other ways that you can reduce land degradation and improve soil and pasture plant conditions; these are also covered further in this book.

Horses, if left to their own devices, will tend to create bare areas more readily than other grazing animals.

Too much water

In some areas or times of year, it is mud and not dust that is more of an issue. Mud occurs when there is too much grazing or standing pressure on land that is wet. It may be because the soil is able to retain lots of moisture, or it may be that the underlying soil, which should allow drainage, has become compacted and therefore, the land retains the moisture. The water is not draining, either because the subsoil will not allow it, or the area that the water usually moves on to is also saturated. Sandy soils tend to drain better, whereas clay/loam soils often become muddy in wet conditions. Manure can hold a great deal of water, so even if you have sandy soils, muddy conditions can be created if there is lots of manure around the gateway.

When horses spend too much time on land in this condition, they churn up the soil and create a 'soup' of mud, water and manure. Although some plants thrive in prolonged wet conditions, most pasture plants can not. Good drainage is essential for pasture health; a wet, poorly drained paddock will have reduced pasture growth.

If you have waterlogged soil, it will need careful management if it is not to become worse. Horses must not be allowed to remain in wet paddocks. Allowing horses to graze on waterlogged soil will lead initially to muddy conditions as the land undergoes 'pugging' (the hooves of animals push into wet soil and leave hoof shaped indentations, water then collects in these indentations). Heavy animals walking or standing on wet soil also create more compaction, as air is pushed out of the soil. If and when the soil dries out, it will be much more compacted than before and is even more likely to become waterlogged next time it rains.

Allowing horses to graze on waterlogged soil will lead initially to muddy conditions as the land undergoes 'pugging'.

In addition to making caring for horses harder work, mud has many disadvantages for horse health:

- It creates a breeding ground for insects such as mosquitoes and midges.

- It can cause injuries - horses can slip or fall.

- Feeding horses on muddy ground leads to soil/sand ingestion.

- Bacteria and fungi proliferate in muddy conditions; these cause skin and hoof conditions in horses such as mud fever, greasy heel, rain scald and thrush.

Horses do not voluntarily stand around in muddy conditions for long periods of time; in a natural living situation horses can usually take themselves to higher, drier ground when they wish. When domestic horses are fastened in a paddock, they cannot do so, so it is up to you to make sure horses do not have to stand around in wet soil.

Again, **The Equicentral System** can work especially well for wet conditions – presuming that the surfaced holding yard is built on a high and dry part of the land, as the horses will take *themselves* to higher, drier ground - as they would in a natural living situation (see Appendix: *The Equicentral System*).

Water that cannot be absorbed should be channelled so that it runs *gradually* towards any dams, streams and other waterways. Aim to channel water *along* contours whenever possible, gradually taking water down to a lower level rather than allowing it to rush downhill at speed. This can be difficult as land is often subdivided without taking contours into consideration, both within a property and between properties.

Horses do not voluntarily stand around in muddy conditions for long periods of time.

Water can be channelled and slowed down using materials placed on the surface of the land such as logs, bales of hay/straw, branches, bundles of sticks or mounded earth; these are called swales (see the section *Using swales*).

Small areas, or even large areas on a larger property, can be fenced off and allowed to become wetland or to regenerate as wetland if that is what it was before. If the waterlogged area is near a water course, then this area, as well as the water course itself, should be fenced off from grazing animals.

Certain water loving trees and other plants can be planted in wet areas (once fenced off) to create habitat for wildlife. Planting the right trees around the edge of each paddock will help to dry paddocks out. Check which species are local to the area and which ones do well in wet soil with someone who has local environmental knowledge, such as a land/soil conservation group.

If rejuvenating wet areas that have become degraded, in order to use for grazing at an appropriate time of year, choose grasses that have a tendency to mat, as this resists the damage that hooves cause. Only allow grazing by horses in these areas at times that will cause minimal damage, at other times mow or use other animals such as sheep and goats.

If fencing a block of land from the start, divide the block into dry paddocks and wet paddocks so that horses can be rotated around the land, keeping them in dry paddocks in wetter seasons and vice versa. If the land is already fenced and the fences are in the wrong position, you can use electric fencing to separate wet areas from dry areas. It is not always possible to know which areas do what at what time of year until you have owned or managed the land through at least one cycle of seasons,. Initially, using temporary electric fences can save you from making expensive mistakes.

If *all* of the land tends to become wet at certain times of the year, then you will need to either arrange to keep your animals somewhere else for this period of the year, or confine horses to surfaced holding yards until the land has dried out, presuming that there is enough high and dry land on the property to build such facilities (see the section *The importance of surfaced holding yards*).

If the waterlogged area is a gateway or other high traffic area, this can be improved by surfacing with gravel (stones, etc.), woodchips or even man made products such as geotextiles or plastic/rubber grid. If using gravel, it will need to be at least 10cm (4ins) thick, preferably more. Try to apply this surface before the mud occurs e.g. at a dry time of year and if possible, put a membrane down first. Make sure that any water that runs over the area is diverted and pick up manure from this area on a regular basis. Reduce the amount of time that horses stand around in gateways (see the section *The importance of surfaced holding yards* and Appendix: *The Equicentral System*).

There can be excessive water around buildings such as stables and barns from runoff from roofs/gutters etc. This often adds to the issue muddy/waterlogged conditions in these high traffic areas. Think about collecting this runoff in tanks for later use, rather than adding to an existing problem.

A gateway or other high traffic area can be improved by surfacing with gravel (stones, etc.), woodchips or even man made products such as geotextiles or plastic/rubber grid.

Not enough water

Dry conditions bring their own problems. Dry conditions are often coupled with hot weather therefore the plants become increasingly dehydrated and will shatter and eventually die if grazing animals are allowed to overgraze them at this time. When the soil is dry, it is also hard and dusty. Hard ground jars the legs of horses, whilst the dust causes eye and respiratory problems in horses.

Prolonged dry conditions mean drought and during drought, plants stop growing due to having no water. Aim to *not* reduce the plants below about 5cm (2ins) in height in order to minimise land degradation; this will help to protect soil from harsh, dry, hot weather. It will also mean that there is enough leaf area remaining for the plants to make a rapid recovery when the rains return. Caring for pasture during drought involves reducing grazing pressure as much as possible during the drought and for some time afterwards in order to allow pastures to fully recover.

Prolonged dry conditions mean drought and during drought, plants stop growing due to having no water.

During drought, a farmer may decide to reduce stock numbers. A horse owner may not want or be able to do this, because they typically do not keep horses for the same reasons that a farmer does. In many cases, depending on the amount of horses compared to the amount of land, a drought may mean that horses have to be removed from the land completely and kept in surfaced holding yards, otherwise land degradation will occur. Remember, if there is no grazing available, you will need to remove the horse from the area to protect the remaining plants and soil, otherwise you will further compact and damage the soil, making recovery much more difficult. A 'sacrifice area' can also be used (see the section **'Sacrifice areas'**).

During drought, grazing animals will have to be fed hay and, if necessary, hard feed (concentrates). Feeding horses during a drought becomes increasingly problematic as the drought goes on; all types of horse feed become increasingly expensive and may eventually be virtually impossible to source. Grass hay usually runs out sooner than legume hay, such as Lucerne/Alfalfa, and concentrates. This is because grass hay is the hardest to grow during a drought. Legume hays and grains for concentrates are usually grown under irrigation, with water usually taken from underground sources. Ordinary grass hay paddocks are not usually set up for irrigation, so less grass hay is made during a drought.

Feeding horses during a drought sometimes involves having to buy hay from sources that you would not normally, due to the risk of infecting your property with foreign weeds. Therefore feeding hay in a surfaced holding yard, stable or a 'sacrifice area' becomes even more important.

In a drought, feeding hay in a surfaced holding yard, stable or a 'sacrifice area' becomes even more important.

If you have no choice but to feed weedy hay due to this being all that is available, always feed it in yards or 'sacrifice areas'. This way, you will more easily see any weeds that attempt to become established after the drought has broken, and you will be able to deal with them before they get established. Also, the weeds will not be integrated into your good pasture; a situation that is much harder to deal with.

If you live in an area that is likely to be affected by drought, it is a good idea to stock and store hay on your land, buying it when it is cheap, abundant and weed-free. This then reduces the effect of a drought for your land, and means that you will not have to rely on pasture as much. Doing this will save you a lot of money in the long run. When a drought starts, aim to top up this reserve if possible; even though the hay may seem expensive, it will be nothing compared to what it may end up, as the price of hay can increase by ten times its usual value during a drought. You can always sell it to other horse owners in the unlikely event that you end up with too much.

Once the drought is over, resist the temptation to turn horses out until the pasture has fully recovered, and start with just a few hours at a time.

Once the drought is over, resist the temptation to turn horses out until the pasture has fully recovered.

Due to being stressed by the drought and accumulating sugars and starches as a coping mechanism, this grass will be *very* high in sugars and starches until it has had time to recover and grow properly; using up these energy reserves to grow new leaves. Horses will gorge on this grass because they will not have had access to fresh grass for quite a while. Many cases of laminitis occur following a drought due to the combination of the plants being stressed and the horses being desperate for green feed.

'Horse-sick' pasture

See the section ***The 'dunging behaviour' of horses***.

Weeds

What is a weed? A plant that is in the wrong place at the wrong time would be one way of defining a weed. The term 'weed' covers a huge range of plants and what is classed as a weed to one group of people may be a beneficial herb or useful plant to another. A particular garden plant may be regarded as beautiful by gardeners, but if it invades grassland, it would be regarded as a weed by farmers. Some grasses that are regarded as weeds by dairy farmers, because they do not yield enough sugar and starch, would be regarded as good horse pasture grasses by some horse managers, due to being safer for recreational horses to eat. Some plants previously thought of as weeds by horse owners may in fact be beneficial in a biodiverse grazing environment.

Weeds tend to be strong competitors, moving in when conditions favour them, especially when the soil is bare or in poor condition. They range from plants which are simply unwanted, to noxious plants that cause huge problems for people, animals and the environment. Some weeds are plants that simply have very little or no feed value; they reduce pasture yield by taking up space, and compete with useful pasture plants for moisture, sunlight and nutrients. These weeds are still a big problem as they rob you of feed and also cost time and money to control.

Some weeds are highly invasive of both native and agricultural land. These types of weeds often have vigorous characteristics which is why they are so important in terms of controlling them. Some of these weeds may be also poisonous to horses (see the section *Poisonous trees and plants*).

Certain characteristics are common amongst weeds and help ensure their survival. Weeds tend to possess one or more of the following:

- Abundant seed production.
- Rapid population establishment.
- Seed dormancy.
- Long-term survival of buried seed.
- Adaptation for spread.
- Presence of vegetative reproductive structures.
- Ability to occupy sites disturbed by human activities.
- High NSC content to enable them to grow quickly.

However, weeds can actually have certain beneficial qualities:

- They can help to stabilise and protect bare soil.

69

- Their strong tap roots can 'mine' deep lying mineral sources and bring them to the surface for the benefit of soil microbes; therefore they can help to correct soil mineral imbalances.
- These tap roots also help to break up compacted soil.
- Some weeds are known to have health benefits, for example Camomile.
- They increase biodiversity
- They can be stores of nectar for bees and food for other animals.
- To the trained eye, weeds are indicators of the condition of the soil.

A healthy vegetative cover is essential for good soil. Even *certain* weeds can be better than nothing and indeed they do many of the things that preferred plants do, apart from providing safe nutritious fodder for stock. Therefore, weeds that are not dangerous to stock should only be removed when you are ready and able to replace them with a preferred species if the alternative is bare soil. In fact, out-competing weeds with a preferred species is often better than immediate removal in many cases, so that the land is never completely bare. You should, however, prevent them from producing seed and spreading further.

What is classed as a weed to one group of people may be a beneficial herb or useful plant to another. Patterson's Curse, invasive weed in Australia, protected species in Spain.

For example, in the case of a steep bank into a river, eradicating weeds without first growing more favourable plants may result in serious erosion. If these particular weeds need sunlight for survival, then a better plan of attack would be to plant trees (that are natural to the area) on the banks. Eventually the weeds will die out and the tree roots will take over the job of holding the river banks together. You may have to cut back some of the weeds in order to give your preferred plants a head start, but it always pays to always get good *local* advice. If the weeds in question are noxious, then this approach will not be fast enough and it is even more important to seek expert local advice.

Weed infestation can be seen as green degradation on a horse property. Weeds can become a problem on a horse property, especially if the land has 'horse-sick' pasture, bare compacted/degraded areas and nutrient depleted or imbalanced soils. Even if a property has good soil, you may still have weed infestations. Many owners ignore these weed infestations as the horses can still eat grass between the weeds; however it is just as important that owners of these properties manage their land to maximise pasture and prevent environmental damage and health issues for their horses. These weed infested properties are an eyesore and strengthen cases for increased legislation against the horse community.

Control of weeds

There are various strategies for dealing with weeds, ranging from minimum to maximum intervention, and often a combination of more than one strategy works best. With many weeds, it is very important that they are not allowed to set seed, as this means that there will be even more of them next year. It always pays to seek local advice when tackling weeds.

You need to know what plants you have on your land so that you know what to do about them; knowing your weeds will enable you to exploit their weaknesses. Your local authority may have a weeds officer who will help you to identify weeds and give advice about how to control them. Your local land/soil conservation group may also be able to help. Don't be nervous about contacting these experts, they are there to help and will provide lots of invaluable advice and assistance. In addition, your local library will have books and your local agriculture department will have fact sheets that enable you to identify weeds on your land and give you information on how to deal with them.

How a plant is classified depends on where you live; some areas take certain weeds more seriously than others. Your local authority and/or agricultural department usually decide how serious a threat a certain plant represents, and will class it accordingly. What you are expected to do about a particular plant depends on how that plant is classified.

Even if a property has good soil, you may still have weed infestations. Many owners ignore these weed infestations as the horses can still eat grass between the weeds.

As a landowner/manager, you are expected to control weeds on your land, and usually on the area between your land and the road (the 'nature strip') as well. There may even be a legal obligation in the case of some noxious weeds. In most areas, there are 'statutory' requirements to ensure that certain weeds do not spread and threaten the wellbeing of other landowners, the community in general and the health of livestock. Check with your local authority/agriculture department about regulations that will affect you and your land and to find out what is expected of you in your area.

Weeds arrive on land in various ways; they blow in from neighbouring properties and the roadside, they are carried in by birds (in their droppings), they come in with loads of soil and gravel and they stick to other objects such as car tyres and animals. They grow creepers, both under and above ground, invade from neighbouring properties, come in with the droppings of new or visiting horses and they come in with hay and other fodder that is brought on to the land. So, the first line of defence is to try to prevent weeds from getting on to your land in the first place.

Preventative measures

- Avoid bare soil on your land; bare soil is an invitation for weeds to become established. Good pasture management *is the best defence* against weeds, so don't allow your land to become degraded and cover all bare areas in paddocks, laneways and yards. Utilise good grazing management systems so that pasture is healthy and vigorous. Prevention is better than cure and the best prevention against weeds is to have vigorous healthy grass that carpets the ground.

- Be very careful when purchasing hay. If you buy directly from a grower, you can find out which weeds are a problem in that area and check that the property is not infested. Check the hay before buying by asking to see a bale opened. Understandably, the farmer will only be willing to do this if you are about to buy a reasonable amount. If you are buying from a feed store or hay merchant, ask if they have checked the quality of the hay and whether they are prepared to guarantee the quality. With all purchased hay (as opposed to home grown), it is advisable to feed it in yards or stables rather than in paddocks. You will be able to monitor and control any weeds that crop up in these areas far easier than if they get into your paddocks. If the hay is weed infested, some seeds will still get into your paddocks via the manure that horses drop while grazing, however this will be less than if you feed them the infested hay in the paddock.

- Make sure you know what is growing in your garden. Be aware that many particularly invasive weeds were initially introduced as garden plants.

- Make sure any contractors wash their equipment prior to arriving on your land.

- If anyone visits your land with a horse, make sure they take their horse manure home with them. Also, beware of the hay that they are using. If you have other people's horses staying on your land, temporarily or permanently in the case of a livery yard (boarding/agistment facility), then consider bulk buying hay to sell to owners rather than allowing people to bring hay on to your land. Remember, it is your land and you may have to deal with the problems of weeds long after they have gone. Alternatively, have strict rules about where hay can and cannot be fed on the land. Although this will not totally prevent weeds from getting into your pasture, it will help a lot.

- Establish multi-storey shelter belts around your land to reduce the number of weed seeds blowing in from neighbouring properties and the roadside. These shelter belts will have many advantages above and beyond keeping weeds off your land, so they are well worth creating (see the section **Windbreaks and firebreaks**).

- Make sure that pasture seed is certified as weed free (beware of buying cheap seeds).

Avoid bare soil on your land; bare soil is an invitation for weeds to become established.

Mechanical control

- Mowing *usually* favours grass plants rather than weeds and can reduce the seed setting of some weeds if done at the right time e.g. before the seeds have ripened, otherwise you may simply be spreading seeds. Mowing is a good strategy for controlling tall weed plants, but is not usually effective for low growing plants as the blades will simply pass over the top of the plant. Also, these low growing weeds will then receive more sunlight as a result of the mower cutting back taller plants and will grow more vigorously. These low growing weed plants *may* need to be treated with chemicals.

- Some weeds can be thrashed by hand with a whipper-snipper (brush cutter) or a scythe. Some plants such as Stinging Nettles wilt when cut and horses enjoy eating them.

- Hand pulling/hoeing by hand. This is very hard work but is a good strategy for small weed infestations and for keeping on top of paddocks or areas that have previously been treated more aggressively. It must be done regularly to be effective. Rather than just pulling weeds out, always replace with seeds from a preferred species. Some plants, such as Ragwort and Fireweed, are poisonous to hand pull as the toxins can be absorbed through the skin. Always wear gloves and in some cases a mask.

Some plants, such as Ragwort and Fireweed, are poisonous to hand pull as the toxins can be absorbed through the skin.

Organic control

- Flame and steam can be used for weeding. Both methods are ecologically sound. Obviously, using flame in hot dry weather should be avoided.

- Cross grazing with other animals can be very useful, especially sheep and goats which are more resistant to certain poisons such as those found in Fireweed and its relative Ragwort. These animals are also invaluable for controlling weeds such as Blackberry, Gorse (Furze) and Patterson's Curse.

- Paddocks can be intensively grazed (usually with other animal species such as sheep/goats or cattle) prior to planting new pasture. This will depend on the species of plants present and should be done in the summer/dry period before you plan to plant the new pasture. The area may need to be ripped/aerated before planting (see the section *Creating suitable pasture*).

- Weeds can sometimes be out competed with other more desirable species; healthy grasses are one of the best defences against weeds.

- Mulching can be used to smother weeds in areas that you plan to replant later, or that need the protection of mulch permanently, such as under trees and pathways/laneways. A huge variety of materials can be used as mulch, such as carpet, newspaper/cardboard, woodchips, stone chips/gravel etc. (see the section *Mulching*).

- Weeds can be good indicators of soil conditions. So much so that eradicating some weeds is simply a matter of changing the soil conditions in one way or another. For example, some weeds like acidic soil and the addition of dolomite can help to get rid of them. Weeds thrive simply because the soil conditions are right for them at a certain time and are also nature's way of trying to provide soil cover at all times. Some weeds may even have some benefit for short periods, for example, they may have long tap roots which are able to penetrate compacted soil and provide initial aeration. These long taproots are also able to bring nutrients to the surface from deeper layers in the soil. Weeds may also hold soil together, which is better than it being allowed to wash or blow away. This does not mean that it is fine to have lots of weeds, just that weeds are usually preferable to bare soil; however the aim should be to replace those weeds with preferable plants as soon as possible.

- Some weeds can be controlled biologically by using the weed's natural enemies such as specific weevils or mites. Speak to your local agriculture department for more advice.

- Experiments have shown that weed seeds pass through the digestive system of horses relatively unscathed, compared to other animals where they do not survive. Keeping chickens and ducks and allowing them to free range and 'sort' through manure for seeds helps to reduce the number of seeds that get into the ground.

Chemical control

- Most horse owners prefer not to use herbicides on their land and, in many cases, they are not necessary. Other methods such as those outlined above can be used instead. However, sometimes using herbicides to gain a quick advantage over weeds can be a good approach, if the plan is to introduce better, more organic, management in future. Always adhere to the instructions for use that are supplied with chemicals. The following points are some general tips about using herbicides.

- Certain chemicals (such as Glyphosate – 'Round Up' etc.) will kill every plant it comes into contact with, whereas others are more selective. For example, a broadleaf spray will kill clovers along with broadleaf weeds, but will leave grasses intact.

Keeping chickens and ducks and allowing them to free range and 'sort' through manure for seeds helps to reduce the number of seeds that get into the ground.

- When weeds are killed with chemicals, the soil is left bare until grass grows again, so it needs to be carried out at just the right time, e.g. not at the beginning of a hot, dry summer. You may need to seed the area with desirable pasture species after spraying, otherwise more weeds will grow in the area that you have sprayed.

- Ploughing up such a pasture will not get rid of the weeds, because there will be many seeds already in the soil. In fact, ploughing will cause many new weeds to germinate. This strategy is actually deliberately carried out by some land managers. Once the weeds have geminated, the whole area is then sprayed with herbicides, more than once if necessary, before ultimately being seeded with pasture.

- Plants should usually only be sprayed when green and growing, and prior to setting seed for best results. Spray on still days when at least 12 hours of dry weather is expected. Follow-up spraying may need to be carried out some weeks later to get the weeds that were missed the first time.

- Do not spray within at least 100m (330ft) of a waterway (including dams/ponds etc.) unless you are using a product that is safe for aquatic life. Check the instructions.

- If using a contractor to spray your weeds, make sure they are experienced, use the correct chemicals and equipment, and have the correct operating certificates if necessary. Your local agriculture department should be able to recommend contractors.

- Animals will usually need to be removed from an area that is sprayed for a specified time (follow the instructions). Spraying can make weeds more palatable for a period of time after, resulting in poisoning if the weeds are of certain types.

- Always store herbicides in a locked storeroom.

Whatever method of weed control you use, weed seeds can remain in the soil for many years even after you have removed all the weed plants. These seeds will be waiting for their chance to re-establish themselves. Vigilance is the key and you need to move swiftly on any new outbreaks. In the case of weeds, the old saying 'a stitch in time saves nine' is very relevant.

—

Remember, healthy vigorous pasture plants are the best defence against weeds. The saying 'prevention is better than cure' could not be more true for weeds.

—

Animals will usually need to be removed from an area that is sprayed for a specified time.

Salinity

Salinity is a problem in certain parts of the world. It arises in two ways; from irrigation and from land clearing. Irrigation salinity is caused by salt being deposited on the soil by irrigation water and this salt then accumulates over time. Dryland salinity is a result of wholesale clearing of trees. Trees have deep roots and when they are removed and replaced with shallow rooted plants, such as crops and pasture, the ground water rises to the surface of the soil, bringing with it salt that is meant to be underground. This salt may then also be washed into waterways, causing problems beyond the immediate area. Salt deposits on the land cause plants that are not salt tolerant to die and plants that *are* salt tolerant to invade. Also, rising water tables can cause waterlogging, which in turn causes plants that cannot cope with wet roots to die.

Once affected, land that has salinity is difficult to manage. Remedies include planting trees, planting salt tolerant bushes and pasture plants, not using the land when wet, or permanently fencing off areas and re-vegetating with trees and bushes that are native to the area (see your local authority). Your local Agriculture Department will be able to give you invaluable advice that applies to your locality in salinity affected regions.

Turning land degradation around

Mulching

The term mulching has a variety of meanings, but in this case involves covering bare/compacted soil to protect it and to provide a medium for new plants to grow. Remember – you should aim to *never* have bare soil (see the section **Bare/compacted soil and erosion**).

- When an area of bare soil is covered in mulch, it is protected from the drying effects of the wind. Soil that would otherwise blow away because it is dry and has no plants to bind it, now stays on the land.

- Water that arrives in that area is slowed down and held in the mulch, helping the mulch to decompose *and* soaking into the soil more easily, thus improving the compacted soil.

- Mulch can be used to 'smother' weeds in some cases. See the section **Mulching with round bales**).

- Mulch provides a cushioning layer between heavy objects, such as heavy machinery and horses, and the soil; although in order for plants to grow back in that area, the effects of heavy machinery and horses will need to be temporarily removed.

- If necessary, even while a paddock is in use, bare areas can be temporarily fenced off with electrified tape and then mulched. This allows the area to become established with new grass, without having to cope with pressure from horses at the same time.

- As the mulch decomposes, it provides habitat for numerous species of insects, bacteria and fungi that are beneficial to the soil. This starts the process of reintroducing beneficial organisms back into the soil.

- The decomposing mulch provides a medium for vegetation such as pasture plants to become established again. Handfuls of desirable pasture seeds can be thrown into the mulch once it starts to decompose.

- Mulching can reap dividends with little or no extra outlay and any bare areas on a horse property should be covered with some form of mulch. Essentially, mulch can be created from any form of organic matter; common materials on a horse property include composted manure, shavings (or other types of used bedding), shredded paper or old hay/straw. Lawn mower clippings can also be used as long as horses do not have access to them until they have decomposed (see the section **A word about lawn mower clippings** for an explanation of why finely chopped grass can be dangerous for horses). Grass seeds can then be thrown into the mulch after rainfall.

- Composted manure should always be covered with another form of mulch. In dry weather, it will dry out too quickly and blow away before it has chance to do any good and, in wet weather, it will soak up too much water and become a 'soup' that washes away.

The decomposing mulch provides a medium for vegetation such as pasture plants to become established again.

- Avoid putting manure in a gateway - there is usually plenty there already. Only mulch a gateway if you plan to implement **The Equicentral System,** which means that horses will never stand around in the gateway again. Otherwise, the gateway needs surfacing so that it does not become muddy again.

- Make sure you wait until any new plants are well-established (do the pull and twist test first – see the section *Creating suitable pasture*), before allowing horses to graze the area. Any weeds that do grow in the mulch can be easily pulled out of the mulch *before* they become established.

Mulching with round bales

A great way to mulch an area when grass hay is cheap and plentiful is to feed clean (from weeds) grass hay round bales on bare areas to groups of four or five horses at a time (if there are more horses in the herd then have multiple round bales). The horses will devour and spread the hay out to approximately a 15m/50ft circle, mixing it with their manure and urine. **This process will take about four to five days:**

- It needs to be four or five horses to each round bale, so that the hay is consumed quickly before it goes mouldy from any rainfall.

A great way to mulch an area when grass hay is cheap and plentiful is to feed clean grass hay round bales on bare areas to groups of four or five horses at a time.

- The hay also needs to be of a type of grass that you want on your land.

- The hay must also be low energy if the horses are prone to obesity and/or are not working because it is being fed ad-lib.

Even if it does not rain for a long time, the mulch is still doing a great job protecting the previously bare soil until conditions become right for mulch decomposition and subsequent plant growth.

You will know when the horses have finished eating the edible parts of the bale, as they will start to hang around where they can see you. It is then time to put another bale on the next bare area, or to remove the horses from that area (or temporarily fence it off). During drought or dry times of the year, the mulch will just 'sit' there. The horses will ignore the mulched area while they have fresh hay to eat. Once it starts to rain, horses must be removed from any mulched areas. At this point the mulch starts to break down rapidly; if horses are left in the area, they will be standing in decomposing hay and manure and disturbing any new plants that are trying to get going. The mulched area should be full of seeds from the hay itself or else you can throw a few handfuls of grass seeds across the mulch. After subsequent rainfalls, you will get a beautiful crop of new grass on a previously bare area.

This method works really well on rocky areas that are difficult to get machinery into. In the previous picture, a group of horses were fed round bales throughout the winter (in a subtropical area and therefore a dry time of year) in what was previously problem strip across the top of a paddock. The horses were then removed in the spring, allowing the mulch to decompose in the spring rains and grass to grow in that area. The new soil created in this area was black and beautiful.

People are often reluctant to feed round bales due to the perceived 'wastage'. Do not consider it a waste, think instead of the benefits to the land. In pure economic terms, there is little wastage. A round bale contains the equivalent of approximately ten small square bales of hay, but is often about five or six times the cost of one small square bale. Even if 40% of the hay is not eaten, the 'feed' still costs the same as with small square bales. Another factor to consider is that it is a form of pasture renovation that *anyone* can do. It does not require any special equipment or contractors, or any further outlay. One round bale can be moved about a relatively flat property by two adults by hand (by rolling it). The pasture itself improves when it is ready and when conditions are right. Some people may also be concerned about the horses eating too much if they have access to 'ad-lib' hay. The hay must be a low energy type or have been tested for energy content.

Using swales

Swales are barriers, placed along the contours of the ground, that slow water and give it more opportunity to soak into the soil. They also divert water and are used to best effect on hillsides. As most horse properties are relatively small, using swales, like mulching, is a very easy way of improving your land.

Traditionally, creating swales involved cutting ditches or mounding earth, but you can achieve the same result by simply placing items such as cut and laid vegetation, bundles of branches, old hay/straw bales, logs etc. on the ground. If the area in question is not too near a waterway, even manure can be used. Loose soil and other organic matter such as leaves will build up behind the swale and this organic matter decomposes and creates soil for new plants to grow. You can even place manure on high side of the swale to add organic matter and nutrients.

Swales can be made by simply placing some form of barrier on the ground.

This is the same area some weeks later.

Simply place swales on the ground in areas where the water moves fastest; usually on hillsides. Pioneer plants, sometimes considered weeds, may grow in the first year and will be replaced by perennial grasses in time. You can speed up this process by throwing a handful of desirable pasture seeds over the swale.

The importance of surfaced holding yards

On most horse properties, surfaced holding yards with shelters, or stables with attached outdoor yards are required so that horses can be safely confined when necessary. Initially, a confinement area can be a 'sacrifice area' (see below); however it is a good idea to aim to build surfaced holding yards that also have shelter. If you have extreme weather conditions, you will need these as soon as possible. These yards will be invaluable when it is too wet, too dry etc.

—

See the third book in this series *The Equicentral System Series Book 3 – Horse Property Planning and Development* for information about constructing surfaced holding yards.

—

'Sacrifice areas'

The term 'sacrifice area' means that part of the land is 'sacrificed' so that other areas have time to rest and recuperate. This would involve putting the horses in one paddock, or part of a paddock by using electric fencing, and allowing that area to be degraded (due to the high level of use), but keeping the majority of the land safe from degradation. **There are some things to keep in mind:**

- During drought conditions 'sacrifice areas' can be used to great effect if you 'mulch' the area at the same time, which may mean that the area comes out of the drought better than before (see the section *Mulching*).

- Only 'sacrifice' land if there are no other areas that could be used instead e.g. any land that already has 'hard standing' in place, such as an old farm yard.

- Sacrifice areas do not work well in wet conditions; the horses will quickly be standing in deep mud and you will have associated skin conditions and extensive soil degradation. In wet conditions, it is imperative that you try to use hard standing if you do not have a purpose built surfaced holding yard/s.

- As long as it is not too wet, an area of land can be used that is earmarked for pasture renovation later on.

- Sacrifice areas should at best be considered a short term solution whilst more permanent holding yard structures are constructed

The term 'sacrifice area' means that part of the land is 'sacrificed' so that other areas have time to rest and recuperate.

Soil, pasture and grazing management

In an ideal world, most horse owners would like to have healthy pastures providing the correct nutrients on which to graze their horses all year round. Unfortunately, the reality for many horse owners is far from ideal. In the naturally-living situation, horses would have favoured grazing areas depending on the season, although, especially in winter, these may not be very nutritious, but would enable them to consume fibre of some description. Many horse owners have to use the same land all year round. However, it is possible to improve and maintain your pastures to an optimum level with a little applied knowledge, enabling you to graze your horses efficiently for as long as possible. By using this knowledge, you will save time and money, keep your horses healthy and content *and* be improving the environment. It is one of those rare win-win situations.

Pasture management is all about trying to provide as close to 'natural' living conditions for your horses as possible, e.g. grassy paddocks with water, shade and shelter, and other horses for companionship. In wider environmental terms grassland management is also about maintaining or improving the condition of the pasture rather than allowing it to decline.

On any horse property, there will be booms and slumps in terms of the amount of feed available during the year, and from year to year. Also, the less available land there is in comparison to the amount of horses, the more intensive the management system will need to be. Even on a well-managed horse property, horses may, at certain times of the year, need to be confined either for their own sake, or for the sake of the land.

As a horse owner *and* land manager, you must also acknowledge that it is not possible to keep domestic horses 100% 'naturally'. Unless you own several thousand acres, compromises have to be made. You cannot compromise the environment, because without a fully functioning and healthy ecosystem, sustainability is not achieved and problems start to appear. So, aim for a healthy ecosystem because that will in turn take care of you, your family and your animals.

It is possible to *manage* your horses so that they still get to behave relatively 'naturally' for most or even all of the time. This is where pasture management comes in. Pasture management involves a twofold approach, firstly, utilizing grazing management strategies and secondly, carrying out pasture maintenance.

With correct management, it is possible to have good pasture. By using grazing management systems, horses can be persuaded to eat more evenly and, by using good pasture management strategies such as mulching, mowing and harrowing, your paddocks will be grazed more efficiently. The detrimental effects of the 'dunging behaviour' of horses will be reduced and your horses will have more grass to eat. The grass will be less stressed and will therefore be healthier, which in turn equals healthier horses. The environment will benefit from less or no pollutants reaching the waterways and you will benefit financially, both due to money saved that would otherwise be spent on feed, as well as due to your property increasing in value. Your own lifestyle will also be improved by living in an aesthetically pleasing and healthier environment.

At any given time of the year you will tend to have either too much or too little pasture, the land will be too wet or too dry and your horse/s will either be gaining weight or losing weight.

Understanding soil

Good pasture starts with healthy soil; well-balanced soil will reduce the need for adding various supplements (minerals etc.) to your horse's diet, as correctly balanced plants will do the job for you. Healthy plants help create healthy soil and vice versa.

Good soil has a balance of minerals and trace elements; it contains microorganisms and organic matter, it contains air and water and provides habitat for earthworms and insects etc., many of which in turn further improve the soil.

Soil has different layers, known as profiles, and these layers have different names and functions. The soil we generally see is called topsoil; this is the layer on the very top and it varies in thickness from a few centimetres (1ins-2ins) to about a metre (3ft). The layers underneath the topsoil are known as subsoils, which in turn are on top of weathering patent rock and beneath that there is bedrock. These layers vary in thickness and type, giving a variety of soil types in different areas. Even within a property boundary there may be more than one type of soil.

The main nutrients that soil provides to plants are Nitrogen (N), Phosphorus (P), Potassium (K), and Sulphur (S). Collectively, these are often referred to as NPKS.

Other elements that soil needs to provide to plants are Calcium (Ca) and Magnesium (Mg). Plants also require trace amounts of elements such as Copper (Cu), Manganese (Mn), Zinc (Zn), Iron (Fe), Aluminium (Al), Boron (B), Molybdenum (Mo), Cobalt (Co) and Selenium (Se).

All of these elements need to be present in the correct amounts otherwise an imbalance is caused which will result in poor plant health or the death of plants. If certain nutrients are missing or are too low in quantity, then this may result in other nutrients being unable to operate properly – even though they are present in the soil. Nutrients become depleted from soil in many ways including leaching (washing out), hay making (if it is sold off a property), collecting and selling manure rather than spreading it (as with hay, nutrients leave the property), soil exposure (such as when land is ploughed and left uncovered) and soil erosion (when dry soil blows away or wet unprotected soil washes away).

The soil we generally see is called topsoil.

Soil texture and types

In order to improve soil, the soil structure must be improved. This involves increased aeration, water infiltration, organic matter and nutrient utilisation. Good soils tend to have darker topsoil, as the effects of organic matter darkens soil. Soil texture is determined by the size of the mineral particles it contains. Particles range from sand to silt to clay. The amounts in which these particles are present give soil its texture. This texture is very important in the management of land.

Some soils are much more susceptible to land degradation than others. Sandy and clay soils are at opposite ends of a spectrum, and understanding how they behave goes a long way towards understanding their management. Basically, sandy soil will need extra protection from horses and other large grazing animals in dry weather, clay soil will need extra protection in wet weather. Good soil is a combination of sand and clay because then it contains the best of both worlds. These soils are called sandy/loam and loam soils.

Soil structure is a delicate balance that can be hindered by excessive pressure from grazing animals through over grazing, leading to compaction (the soil particles compact together) and pugging. These conditions lead to a reduction in plants. Plants provide organic matter to soil and their roots keep the soil 'open', allowing air and water into the soil – so when they die the soil becomes increasingly degraded. Another practice that degrades soil structure is excessive ploughing and cultivation, which breaks up soil aggregates, exposes soil to the air and disrupts microbes in soil until it eventually becomes a dust.

Soil structure can be improved by adding certain soil additives. However, the soil structure will improve itself over time once air, water and organic matter get to work. Getting this cycle started will involve having soil tests done to find out what deficiencies the soil has and then correcting these deficiencies with soil additives and/or fertiliser in order to get plants started. If the land is compacted, it may need to be loosened up with an aerator implement that cuts into the soil without mixing the layers so that new plants can grow down into the soil. Ironically, even though cultivation should be avoided when possible, it may be the only way of creating the conditions necessary for delicate new plants to get established if the soil is compacted. A subsoil ripper should loosen the subsoil and leave the topsoil and any plants relatively intact. Used correctly, this implement vastly improves soil.

Some soils are actually too high in nutrients, especially if you want to re-establish native pasture. This can be the case in areas that have been used for intensive farming for many years. So, you should decide what type of pasture you would like to establish before you add anything to your soil. If your plan is to encourage or grow native grasses, make sure you indicate this when you send your soil test to a soil testing laboratory.

Difficulties that you may encounter when trying to improve sandy soil are that it is fragile when dry, does not hold moisture very well and nutrients are leached out of it easily. While this means that sandy soils do not tend to get waterlogged, unless the subsoil is very compacted, it means that they dry out too quickly, and that nutrients and fertilisers tend to be leached through them quickly. Some deep sands have no structure at all and getting a cycle of growth started can be difficult. These soils are often so *well* aerated that microbes burn up any organic matter very quickly.

Good soils tend to have darker topsoil, as the effects of organic matter darkens soil.

The aim with sandy soils is to increase their organic material component and soil structure as *quickly* as possible. This can be done by planting deep rooted and quick growing legumes and grasses and then mowing them frequently so that the roots die back and re-grow (as the plant re-grows), rapidly creating organic matter. This method is referred to as 'green manuring', as the plants provide nutrient rich organic matter both through their roots and the leaves and stems that are cut back and fall to the ground to decompose.

Clay soils have their own disadvantages because they hold *too much* water and are therefore prone to waterlogging. When they dry out, they tend to be too dry and crack; they do however retain nutrients better. Like sandy soils, they can also be improved by increasing the organic matter content by introducing plants such as legumes and deep rooted grasses. Once the soil structure is improved due to an increase in organic matter, water can move down, into and through the soil rather than simply across the top.

With *all* soil types, the soil structure is helped by an increase in organic matter because this allows air and water into the soil and increases microbial, fungal and plant root activity. This organic matter keeps the soil 'open'. Once correct conditions are in place, the roots of plants can utilise any nutrients in the soil, sometimes from deep below the surface, depending on how deep rooted the plants are. So, with good management, healthy soil will actually keep improving – maintaining an upwards, rather than a downwards spiral.

Soil and plant tissue testing

Soil tests tell you what is and isn't present in soil, as well as how fertile soil is. Soil testing tells you the nature and condition of soils, such as their acidity and salinity. If soil is not correctly balanced, new plants cannot grow well. Money spent on soil tests is a good investment.

Soil testing kits are available from independent laboratories that are in the business of soil testing – rather than selling fertiliser. Fertiliser companies offer soil testing as a service so that they can then recommend which and how much fertiliser you need to buy from them. Some government agricultural departments will also assist you with soil tests. The returned report will detail the level of plant nutrition available and the soil condition; it will recommend types and amounts of fertiliser and soil conditioner needed. Independent laboratories will usually recommend which nutrients are needed, rather than a particular brand of fertiliser which is what a fertiliser company will tend to do. Rural stores which also supply pasture seeds will usually be able to help you both with soil testing and providing any necessary minerals and fertiliser.

Depending on the soil test results, it is often recommended that minerals such as Lime or Dolomite be added to the soil some time before any re-seeding is carried out. Your soil tests results will indicate what is necessary and when.

Taking samples - Take samples from areas that represent the whole paddock. Don't take samples from areas such as under a fence or near tracks, roads or buildings. If the paddock has well-established 'roughs' and 'lawns' (see the section *The 'dunging behaviour' of horses*), you may need to take samples from both areas. Aim to take soil tests once a year or at least every three years from the same spots. Make a map when you take samples so that you can return to the same areas for subsequent tests.

Even if you are improving just one paddock each year, initially taking samples from the whole property will help you to decide which paddock to improve first. It is better to aim for full fertility in one paddock than to go for partial fertility in all of the paddocks.

Plant tissue tests can also be carried out by some laboratories. These can determine the Calcium (Ca):Phosphorus (P) ratio and levels of other nutrients such as total Nitrogen, Potassium, Magnesium, Sodium, Copper, Zinc, Iron, Manganese Cobalt, Boron and Sulphur. These tests tell you what nutrients the plants are actually utilising from the soil. Nutrients taken up by plants are then able to be used by grazing animals once the plant has been digested.

The pH of soil is a measurement of its acidity or alkalinity, and is measured on a scale from 0-14. The pH affects plant growth and it is therefore important that it is correctly balanced. A pH of 0-6 is acidic, pH 7 is neutral and 8-14 is alkaline. Most plants grow best in a pH of 6-7. If the pH is outside this range, plants tend to suffer from certain deficiencies, as pH effects what nutrients are available to plants. Acidic soils are especially detrimental to legumes, as a low pH prevents the formation of nitrogen fixing nodules on their roots.

Soil additives and fertilisers

Soil additives include materials that improve soil condition and structure *and* add nutrients. Certain soil additives can also improve the absorption rate of soil. It is very important that soil tests are carried out first to determine your soils particular failing before adding anything. Soil additives may not be necessary at all, some additives may only need to be applied once, and others more frequently or at least until the soil is healthy and able to sustain itself.

Soil conditioners

Common soil conditioners include Lime/Dolomite, Bentonite and Gypsum. Calcium is required for acidic soils which 'lock' up certain plant nutrients, making them unavailable to pasture plants. Lime and Dolomite are mainly Calcium Carbonate but Dolomite also contains Magnesium Carbonate. Which one you use depends on availability unless your soil lacks Magnesium (quite rare).

Bentonite is used for many applications including sealing dam walls. It is a soil conditioner, a fertiliser, it neutralises acidic soils and reduces groundwater contamination whilst also being non-toxic to livestock and fish. It is particularly useful for sandy soils that do not hold water as it swells when wet and holds many times its own weight in water. Interestingly, Bentonite can be (and is) used as cat litter for its ability to soak up moisture.

Gypsum is an additive for clay soils, as it improves soil texture, drainage and aeration. It is also a fertiliser and is pH neutral. Gypsum is Calcium Sulphate Dihydrate and is useful in soils that need the Sodium:Calcium balance to be restored.

Other soil conditioners include ground dressings such as poultry manure, compost, kelp and fish emulsion. These additives are soil conditioners, because they add organic matter, *and* also fertilisers, because they add nutrients at the same time. Extra care must be taken that these products do not wash into waterways. Animals must be kept off the land that has been treated until the product has been absorbed by the soil. If buying ground dressings, you need to calculate the cost of the product, taking into account the nutrients that they contain. Adding composted manure that is produced on your land is an excellent way of conditioning and fertilising your soil without having to buy-in products.

Fertilisers

In the case of straight fertilisers, you have a choice of using chemical or organic fertilisers, with the preference usually being for organic varieties. Bagged fertilisers add nutrients but not organic matter, which is fine if you get the correct balance of nutrients, as plants will grow quickly and produce their own organic matter.

There is no doubt that fertiliser has a place in agriculture, especially when improving and establishing new pastures. However, once the correct balance of nutrients is in place, it should be possible to reduce fertiliser use unless the pastures are used for cropping (a hay or feed crop etc.) that is sold off the property. Selling a crop means that the nutrients that have been removed will need to be replaced with fertilisation. If the crop is fed to resident animals, the manure should be used as fertiliser to put back some of the nutrients. Fertilising a crop paddock or a pasture with composted horse manure will return much of the Phosphorus and Potassium consumed by horses. Nitrogen may need to be added if the pasture has few or no legumes.

Over fertilisation (organic as well as chemical) leads to excess run off into waterways and causes all sorts of problems to aquatic life. Soluble chemical fertilisers can kill earthworms and destroy microbiological soil life; they also have a high environmental cost as non-renewable natural resources are used to make them.

Green manuring, growing then mowing plants (in particular legumes) to increase organic matter, is yet another way of fertilising land (see the section *Soil texture and types*).

Different areas have different deficiencies in their soils; even within a relatively small area fertiliser requirements can change to a large extent. Soil conditioners and fertilisers should only be applied if soil tests (or tissue tests) indicate there is a need for them, because some nutrients such as Selenium are poisonous if present in high amounts. In addition, adding nutrients that are not needed is not only costly, but causes imbalances to soil and can cause pollution.

Contact your local government agriculture department for their advice and to find out what is likely to be necessary in your area.

There is no doubt that fertiliser has a place in agriculture, especially when improving and establishing new pastures.

Horses and manure

Manure is a valuable bi-product on a horse property, yet it is often considered as a waste or, at best, as a disposal nuisance. With a little thought and planning, it is possible to use this manure advantageously; in fact manure should be viewed as 'black gold' on a horse property.

The main considerations in relation to horses and manure are:

- Making sure that manure (and urine) does not cause pollution problems for humans, horses, and wildlife. In particular, this means making sure that it does not enter the waterways.

- Managing manure so that equine parasites (worms) are controlled as safely and sustainably as possible.

- Reducing chemical usage e.g. parasitic worming pastes because the parasites are becoming resistant to the chemicals in them.

- Managing manure in such a way that it does not create too much extra labour or expense.

- Complying with any local bylaws about manure use and disposal.

Manure should be viewed as 'black gold' on a horse property.

Some facts and figures about horse manure:

- A normal healthy horse deposits manure about 10-12 times per day. Horses defecate 4-6% of their body weight in dung per day. A confined 500kg (approximately 1,100lb) horse can generate over 10,000kg (approximately 9–11 tons) of manure a year, or twenty times its own weight! Urine adds to these figures, as does bedding.

- Composition of this material varies depending on the type and quantity of bedding used, the age and function of the horse and the type of feed the horse eats. Typically, 1000kg (2,200lb) of fresh horse manure, with bedding, would have a nutrient composition of about 5kg (11lb) of Nitrogen (as N), 2kg (1lb) of Phosphorus (as P_2O_5), and about 6kg (13lb) of Potassium (as K_2O). Manure also contains other valuable trace elements. These valuable nutrients should all be utilised to their full extent.

These valuable nutrients should all be utilised to their full extent.

- Horses excrete many of the nutrients that they consume and well-fed horses produce even higher levels of nutrients in their manure.

- Horse manure is very high in organic matter from the plants that horses eat, for example grass, hay and grain, contain indigestible fibre (cellulose and lignin), which is a form of organic matter. Many soils are low in organic matter, and the high levels of organic matter in horse manure can help to redress this.

- Manure improves soil texture and soil moisture holding characteristics, thereby reducing the need for irrigation. Manure can hold much more than its own weight in water. This fact alone makes it an important aid to retaining moisture in the soil on a horse property; a high priority in dry areas. Even land that tends to be too wet may have dry areas (e.g. higher and dryer parts of the land).

- By selling manure, you are reducing nutrients on the land and reducing the opportunity to increase organic matter in your soil. When you sell manure off the property, other people buy it as it makes a great fertiliser and soil conditioner, yet most horse properties need better soil! Removing manure from the land is in contrast to the natural environment where animals graze plants but also drop manure, which in turn helps to grow more plants.

When you sell manure off the property, other people buy it as it makes a great fertiliser and soil conditioner, yet most horse properties need better soil!

The 'dunging behaviour' of horses

The 'dunging behaviour' of horses is similar to that of other grazing herbivores in that they avoid grazing near any dung piles from their own species, but will graze near that of other species. So, for example, a horse will not graze close to a pile of horse manure but will graze near a cow pat. This is thought to be an innate parasitic worm avoidance strategy.

In addition, horses 'group' their dung in to areas within a pasture (the 'roughs') and graze in others (the 'lawns') this can lead to a situation sometimes termed a 'horse-sick' pasture (see the section ***'Horse-sick' pasture***). In contrast, cows and sheep are not as rigid as to where they drop dung in a pasture.

In a naturally-living situation, horses are not forced to graze over their own manure due to the space that they have available to them. In such a situation, many different species of animal graze the same areas, each grazing near and around each other's manure and taking in each other's parasitic worms which kills the parasites.

In a naturally-living situation many different species of animal graze the same areas, each grazing near and around each other's manure and taking in each other's parasitic worms which kills the parasites.

Parasitic worms are generally 'host specific' meaning that they can only survive when picked up by the host animal with which they have evolved. As equines are not at all related to ruminants (cows, sheep, goats, alpacas etc.), many of their parasites have evolved quite separately, whereas certain ruminant parasites can infect other ruminant animals because the host animals are more closely related. For example, some of the parasites of sheep can also infect alpacas.

Additionally, within this healthy ecosystem, there are many predators that feed on parasitic worms at various stages of their lifecycle. Dung beetles are also prevalent in areas that large grazing herbivores live and these help enormously with reducing parasitic worms by destroying their habitat (see the section ***Dung beetles and other insects***). Therefore, there are a variety of mechanisms that *naturally* reduce parasitic worm burdens in all grazing animals.

In the domestic situation, horses are often confined within the same area for long periods of time. There may not be a healthy ecosystem in place to help control parasitic worms, and indeed other parasites, and so manure management, in effect parasite management, becomes a major issue.

'Horse-sick' pasture

Pastures that have well-marked areas of roughs and lawns are often termed 'horse-sick'. An old fashioned but still used term that generally means that the land has a high level of parasitic worms *and* has well-established roughs and lawns.

Horse 'dunging behaviour' leads to:

- **Progressively less grazing each year** as the ever decreasing lawn areas usually have to provide feed for at least the same amount of horses as the previous year. Without intervention, more and more of the pasture becomes unavailable for grazing over time.

- **An imbalance of nutrients** as horses take from the lawns, by grazing and deposit manure in the roughs. Horse manure is high in nutrients such as nitrogen and, therefore, the roughs receive lots of fertiliser and the lawns do not.

- **A higher number of parasitic worms** on the land because the manure is not being managed properly e.g. picked up on smaller paddocks or harrowed-in on larger paddocks. The number of infective parasite larvae in roughs is many times greater than in the lawns, but these infective larvae do migrate to the lawns, particularly after rainfall, meaning that horses are then infected by them.

- **Monocultures of short, stressed grass and roughs** which contain only long, rank, inedible plants and weeds; especially the types of weeds that thrive on high levels of nutrients. These areas become a 'seed bank' for weeds as the seeds build up and are either deposited in the soil within the rough, or blow across and 'infect' the lawns.

- **A poor impression** because horse-sick land is unsightly and gives the general public a negative view of horsekeeping.

When the dunging behaviour of horses is not managed well it gives the general public a negative view of horsekeeping.

How roughs increase in size

The roughs in a pasture tend to increase in size over time because geldings and mares face in to the centre of a rough to drop manure. Stallions back into it and have a tendency to create a neat pile of manure; this behaviour is scent marking behaviour so that any other stallions in the area know that they are there. The way that mares and geldings drop dung leads to increasingly larger roughs over time.

Why we have to manage this behaviour

Horse 'dunging behaviour', therefore, is something that has to be managed in the domestic situation otherwise horses will render a pasture horse-sick and increase their parasitic worm burden.

Manure management strategies

There are various strategies for dealing with pasture manure *and* manure that is deposited in areas that horses are confined such as stables/surfaced holding yards etc. Pasture manure, in particular, has to be managed in such a way that it does not increase cases of parasitic worms in horses.

Pasture manure, in particular, has to be managed in such a way that it does not increase cases of parasitic worms in horses.

Managing pasture manure and parasitic worms

Grazing animals excrete parasitic worm eggs in their dung, and these hatch out to become larvae which are then picked up and eaten by the grazing animal (or others in the vicinity).

The major parasitic worms that affect equines are the large and small strongyles, stomach worms, small stomach worms (hairworms), tapeworms, roundworms and pinworms. Stomach bots are another type of parasite, being the larvae of a type of fly. Bots are not generally as harmful as parasitic worms, however they still require control.

Chemical parasitic worming pastes (anthelmintics) cannot totally be relied upon to control parasitic worms because the worms are becoming resistant to the chemicals in them. In fact, in time, these products may not work at all. The use and *overuse* of these chemicals is causing problems both for horses *and* for the environment and we must therefore adopt good grazing management and manure management strategies to use *in conjunction* with the careful use of chemical parasitic worming pastes.

Controlling parasitic worms is best done by utilising *various* strategies that tackle the number of parasitic worms in individual horses, e.g. using parasitic worm egg count tests *and* using chemical parasitic worming pastes *as well* as controlling the number of parasitic worms on the land.

—

See our **Equiculture website page** about horses and worms for more information on the subjects of using parasitic worm egg count tests and chemical parasitic worming pastes. There are also links on the page to other websites that have good up to date information. **See www.equiculture.com.au/horses-and worms.html**

—

Manure in paddocks *must* be managed if horses are not to increase their parasitic worm burden and render the land horse-sick. Doing nothing with the manure is not an option. Horse owners often pick up manure from paddocks (see the section *Picking up manure in paddocks*) however it *may* not be necessary to do this if you carry out other parasite control strategies **such as:**

- **Pasture rotation** so that *some* of the parasitic worm larvae die because there are currently no horses in the paddock to eat the plants that they are attached to. See the section *Rotational grazing*.

- **Harrowing** to spread the manure which dries it out. Parasitic worms need moisture in order to survive, so this also kills *some* of them. See the section *Harrowing*. This practice also further exposes them to the effects of strong sunshine and frost, which also kills *some* of them.

- **Cross grazing**, which is using other animals to graze the same area of land, so that *some* of the parasites are picked up by them, rather than by horses, and killed by their digestive systems. See the section *Cross grazing*.

- **Encouraging/introducing dung beetles** which take fresh manure underground. These are the easiest and most sustainable option once established. See the section *Dung beetles and other insects.*

- **Utilising chickens** as they *can* also help with paddock manure (see the section *Chickens and horses*).

Dung beetles and other insects

Dung beetles are a very important part of good horse property management and therefore you need to understand them, so that you encourage them, rather than disrupt their functions.

At certain times of the year, you may notice that piles of manure in the paddock are being broken down, and all that is left is a layer of fibrous material on the surface. You may also see tunnels going down into the ground underneath what is left of the manure. This is usually the work of dung beetles.

There are various types of dung beetles and different countries vary in what is available. It is usually best to have several types as they tend to work at different times of the year, or indeed will work alongside one another for a better result. Dung beetles work on fresh piles of manure, not on old or composted manure, generally flying to a fresh pile of manure at dawn or dusk. They then either burrow down into the soil underneath the manure pile, taking manure with them (borers), or they roll balls of manure to a nearby burrow that they prepared earlier (rollers). A third type of dung beetle are classed as 'pad dwellers', these dung beetles are not as beneficial as they stay in the pad of manure rather than take it underground or roll it away.

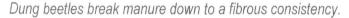

Dung beetles break manure down to a fibrous consistency.

The results of dung beetles (particularly borers and rollers) are hugely advantageous:

- **They dig tunnels into the ground**, improving the soil condition by aerating compacted soil and allowing rain water to penetrate. This also allows the plant roots to penetrate the soil more easily.

- **They create the right conditions for beneficial earthworms** by loosening the soil. Earthworms are also vital to the soil so it is very important that they be encouraged.

- **They clear the ground**, which enables grass to grow in that spot straight away, rather than being trapped under a heavy pile of manure.

- **They remove habitat for flies and parasitic 'worm' larvae**, as these pests need intact piles of fresh manure to survive. The small amount of desiccated manure left on the surface is not a viable habitat for flies and equine parasitic worms – especially in dry weather.

Dung beetles dig tunnels into the ground, improving the soil condition by aerating compacted soil and allowing rain water to penetrate.

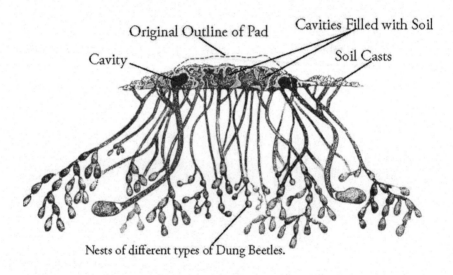

Nests of different types of Dung Beetles.

- **They take the nutrients from manure deep down into the soil to the level of the plant roots**. The nutrients in the manure that they take down are able to be used by the soil and plants, as the process of digestion by the beetles makes it more readily available to them.

- **They reduce and sometimes eliminate the need for manure to be picked up or harrowed**. When dung beetles are active, there is nothing to pick or harrow! Therefore they save you an enormous amount of time.

- **They reduce the effects of horse 'dunging behaviour'** because horses will graze in this area rather than reject it as the manure has been taken underground.

Who would have believed that a small insect could do so much! Take care of these hardworking manure management experts by minimising chemical parasitic worming paste use. Select worming pastes that do not kill dung beetles (at any of their various life stages) and avoiding picking up manure from paddocks or harrowing when dung beetles are working, because this disturbs their habitat. Remember, these guys are the world class experts; nothing that you do with manure comes close so, when they are working, let them get on with it!

It is usually possible to purchase dung beetles, although correct manure management will sometimes encourage them naturally if they are already present in your area.

—

The **Equiculture website page** has links to various websites about dung beetles. See **www.equiculture.com.au/horses-and-dung-beetles.html** where you will find out more about dung beetles in your country, including how you can obtain them and how you can take care of them. Have a look and follow the links.

—

Other beneficial insects

There are other insects that may be beneficial for manure management; some of these may eat the eggs/larvae of parasitic worms and flies, and some help to break down dung. Included are certain staphylinids (rove beetles), spiders, ants and predatory mites, earthworms, millipedes, certain beetles and certain flies.

Good manure and pasture management means that we have to take care of their environment, so that these beneficial organisms can do their work. This involves various strategies including being careful with chemicals, such as those used to kill equine parasitic worms.

Picking up manure in paddocks

Should you pick up manure from your paddocks or leave it? It comes down to whether you rotate your pastures or not (see the sections **Set-stocking** and **Rotational grazing**). If you keep horses permanently in paddocks without rotating them around the available pasture then manure *must* be picked up (unless dung beetles are working). Conversely, it is not *as* necessary to collect manure in paddocks if the horses are rotated around the land. Harrowing and/or Cross grazing will also need to be carried out (see the sections **Harrowing** and **Cross grazing**). Paddock rotation has huge benefits in terms of pasture *and* manure management, not least of all saving you time that would otherwise be spent picking up manure.

Therefore, collecting manure *is* necessary in paddocks that are set-stocked. Manure can either be picked up by hand (e.g. a shovel and wheelbarrow), or machines can be purchased that 'suck', 'sweep' etc. the manure. If you plan to buy such implements, make sure you do your homework first. They still require time and labour e.g. you still have to go out into the paddock with the machine! Ease of use should be very high on the list of priorities, because otherwise you may as well use the inexpensive 'old fashioned' method of a shovel and barrow.

In order to be effective against the formation of roughs and parasitic worm infestation, manure must be removed either daily or every two days at the most, (if other manure management strategies such as paddock rotation, harrowing and cross grazing are not being utilised). If manure is left any longer, the benefits of removing it decrease. The longer that manure is in the paddock, the more opportunity there is for equine parasitic worms to become established.

Manure that is collected from paddocks, stables and surfaced holding yards should be improved in some way and then used or sold (see the sections *Improving manure* and *Using improved manure*).

Manure that is collected from paddocks, stables and surfaced holding yards should be improved in some way and then used or sold.

Managing collected manure

Obviously manure must be collected from areas where horses are confined (such as stables and surfaced holding yards), but some horse owners also pick up manure from paddocks (see the section *Picking up manure in paddocks*).

Improving manure

There are various strategies for improving manure. This section outlines some of them, namely composting, vermicomposting and using chickens.

Composting manure

Composting is a process whereby bacteria and fungi consume oxygen while feeding on organic matter. This results in the release of heat, carbon dioxide and water vapor into the air.

Decomposition of manure starts as soon as it is passed and the rate of decomposition depends on handling and storage methods. Composting manure is a great way to turn a nuisance into an asset, it is a method of *speeding up* the process of decomposition that occurs naturally to organic materials; organic materials can be defined as something that was once alive. By providing the correct conditions, the micro-organisms that decompose organic matter, such as manure, are able to work to their full extent. The end result is a far superior product than fresh manure.

Decomposition of manure and bedding under composting conditions has several benefits:

- **Composted manure is safer to spread on pasture** as it is not as harmful as manure that has not been composted if it enters the waterways.

- **The composting process brings the ratio of carbon to other elements into a balance**, providing plants with nutrients in the absorbable state. All organic materials contain Carbon and Nitrogen, although the ratios vary from one material to the next. An ideal ratio is between 25:1 and 30:1. Horse manure alone is close to an ideal ratio. Grass clippings, for example, have a ratio of 20 to 1 (too much Nitrogen), whereas straw and wood shavings have too much Carbon.

- **Composting makes nutrients such as Phosphorus more available to plants**, as it changes the organic matter into substances that more readily form humus in the soil. About 50% of the Nitrogen in *fresh* manure is soluble and can be washed away by rain whereas after composting about 95% of the Nitrogen is in a more stable form and is able to be released slowly. This slow releasing of nutrients means that nutrients are available to plants (albeit in decreasing amounts) in the seasons after application. With a regular topping up of composted manure, the need for other fertiliser is very low or even eliminated altogether.

Composting manure is a great way to turn a nuisance into an asset, it is a method of speeding up the process of decomposition that occurs naturally to organic materials.

- **Compost applied to the land improves aeration and water retention as well as adding organic matter to the soil**, this means that, in sandy soils, compost helps the sand to hold water (compost can hold almost two times its own weight in water) and, in clay soils, compost loosens the packed clay by opening up pore spaces, allowing water and air into the soil. When soils are able to hold water for longer, they do not dry out as quickly in a dry period.

- **Improved soil feeds and encourages earthworms**, earthworms further improve the soil; therefore an 'upward spiral' occurs, with the soil continuing to benefit.

- **Compost gradually changes the pH** of soils that are either too acidic or too alkaline.

- **The high temperatures achieved when composting kills weed seeds, parasitic worms and pathogens** such as bad bacteria, viruses and fly eggs and larvae.

- **In properly enclosed compost systems, the breeding ground for flies is also reduced** (compared to an open manure pile) further reducing the fly problem.

- **Odours are also reduced**, especially in a covered bay.

- **Composted manure has roughly 50% less volume and weight** than manure that is not composted.

To compost manure, you must keep it moist, and either turn it over several times for one or two months, or establish a system of aeration. Various techniques can improve and hasten the composting process. Processing methods can be kept quite simple, or be quite sophisticated, depending on the desired condition of the end product and the time needed to complete the composting process.

- Air and water are required for decomposition to take place; the pile should therefore be turned at regular intervals. Alternatively, insert some form of perforated piping into the pile to get air into the centre, and consider slated sides for the composting bay for more aeration. The pile should be kept damp but not too wet (the consistency of a wrung out sponge). Covering the pile will help it to stay moist without allowing rainwater to get in and make the pile too wet. Once the bay is full, the front of the pile can be closed in if it was open before.

Covering the pile will help it to stay moist without allowing rainwater to get in and make the pile too wet.

A well-managed pile should decompose in one to three months in the summer and three to six months in winter. For successful composting, the air, moisture and nutrients in the heap must be properly managed. So, ensure that:

- Air can get to the manure pile (especially in the early stages of the process).

- Temperature is maintained at 45°C to 65°C to ensure certain bacteria, disease carrying insects, parasitic worm larvae and any weed seeds are killed.

- Moisture is consistent throughout the manure pile. When adding dry materials such as leaves, grass (lawn) clippings or mouldy hay or straw to the compost pile make sure these items are saturated with water first.

- Odours are controlled by ensuring the compost pile is regularly turned and aerated to prevent any build-up of raw materials, ammonia and other gasses in any one spot.

- There is a blend of raw materials to provide the right ratio of easily composted nutrients as opposed to slow composting woody fibre. It is sometimes necessary to add other ingredients to shorten composting time and minimise unwanted odours, especially if the pile contains a large amount of bedding material.

This is a blown air composting system. Air is blown through the manure pile helping it to decompose.

You can add:

- Household vegetable scraps - all of your food waste should be composted.

- Grass clippings and any other garden waste, but be careful about poisonous plants.

- Leaves and soft branches (branches may require chopping first).

- Mulched/chipped woody material (but not too much).

- Common weeds.

- The quantities and qualities of each mixture will vary, depending on the time of year, Nitrogen ratio in your horses' feed, and successful aeration of your composting pile. A composted pile is hot when it is working and returns to air temperature when it is done. It is damp, crumbly, dark coloured, smells earthy and looks like potting mix.

- Once composted, you are left with a material that is high in organic matter and is similar to potting soil. This material is very valuable and should be used on the land as fertiliser and for improving soils in paddocks and gardens. Of course it also has a high sale value, but why sell such a valuable commodity unless you already have excellent soils?

Composting bays

Bays should be built for efficient composting as, unless a pile is compact, it does not reach the required temperature to compost properly. A pile should be at least 1m (3ft) high to reach high enough temperatures and 1.5sqm to 2sqm (5sqft to 6.5sqft) in area. A compost heap can either be purpose built or made simply from materials such as old hay or straw bales. These bales can be periodically mixed in to the heap, and new ones used to start another compost heap. Other materials that can be used to build a temporary or permanent compost heap include wood pallets, bricks, old tyres, concrete blocks, railway sleepers, steel sheets, timber, concrete etc.

This steel sided composting bay also has a perforated pipe that allows air into the manure pile.

The amount of space required to compost your manure depends on how many horses you have and how they are managed. Other considerations include whether you have access to machinery, or you will you be moving the compost around by hand. A 10m by 10m (33ft by 33ft) pad will comfortably house three compost bays with room to move machinery. This area is large enough to cope with the manure from one to five horses that are kept confined on a regular basis (e.g. in overnight, out through the day). Three bays are ideal so that one can be filling, another composting and the other full of finished composted material that is available to use. A smaller area will be fine for a horse property where less manure is collected (e.g. the horses are at pasture for much of the time) and/or machinery is not used in the process. It may be a good idea to start with a smaller area and add to it if the need arises later on. Make sure you position the pad with this in mind.

A compost heap should be located well away from a water course. There should be a buffer zone of vegetation between it and any waterways. It should also be in the sun. Check with your local authority about regulations for your area. Try to locate storage sites so that filling and emptying the bays will be convenient. It may be possible, to locate a compost bay below the level of the stables so that manure can be tipped into the bay without having to fork it out of the barrow. Covering the heap (with a tarp or piece of old carpet) keeps it moist, reduces the breeding ground for flies and reduces loss of nitrogen to the atmosphere in the form of ammonia. It also prevents water from running through the heap which can contribute to unwanted nutrients entering the waterways.

The ground surface should be non-permeable to prevent leakage of manure into the soil which can then get into the waterways. If possible, a concrete pad should be laid. This is even more important if machinery will be used to shift the compost otherwise the area will quickly become waterlogged and smelly.

Grade the surrounding area to keep surface water from running over or through the manure and then into farm dams, streams or rivers (taking pollutants with it).

Composting worms - vermicomposting

Vermiculture is the practice of breeding composting earthworms – the goal being to increase the number of worms so that they can be sold etc., *vermicomposting* is the process of using composting earthworms to decompose your manure and it utilises the digestive processes of certain earthworms. Once horse manure and food scraps have been through the digestive system of the composting earthworms, it becomes a highly desirable product called 'worm castings'. The earthworms also produce a liquid, called 'worm tea', which is a valuable fertilizer.

- The worms used for this process are a specific type of earthworm (e.g., red worms, tiger worms, red wigglers) that work with other compost organisms such as insects, bacteria and fungi to decompose manure and even some types of bedding. They will also decompose other types of organic matter such as food scraps etc.

The worms used for this process are a specific type of earthworm.

- Why would you use composting earthworms rather than just compost the manure without the worms? By using composting earthworms you eliminate the need to aerate the manure pile. This is quite significant because aerating a manure pile can be the most difficult part of traditional composting.

- To get started you need to acquire some composting earthworms. They are usually sold at garden centres etc. Composting earthworms can double their population every four months under ideal conditions. Therefore, even though you need a lot of worms, you can start with just a kilo or two and they will increase their numbers quickly. In time you may get to the stage where you can sell excess worms (vermiculture), castings and worm tea.

- Before getting started, you will need to do some research – for example you need to position the manure pile (or a 'windrow' – a longer narrow manure pile) correctly. Composting earthworms can dehydrate easily in the heat so you need to make sure that they do not get too hot. Composting earthworms need to be in an environment that has at least 4% moisture (at least as damp as a wrung out sponge). Another thing to keep in mind is that chemical parasitic worming pastes in manure will kill the composting earthworms so you need to be careful with them.

—

Have a look at the **Equiculture website page** for this subject for links to more information. See **www.equiculture.com.au/horses-and-vermicompost.html**

—

Chickens and horses

Chickens and horses have lived side by side for centuries. Chickens can be very beneficial in terms of manure management on a horse property and can help you both with manure that is collected *and* manure that is left in the paddocks:

- They will scratch through manure in the pasture and on a manure pile, eating flies, weed seeds, parasitic worm eggs/larvae.

- While scratching through manure on pasture they spread it. This dries the manure out and reduces the habitat for flies to breed, as well as the habitat for parasites such as equine parasitic worms - both flies and parasitic worms need moisture in order to thrive.

- Chickens are able to carry other functions on a horse property beside manure management such as:

- They eat weeds, food scraps, rodents (dead or alive) and almost anything else you can think of.

- They also keep the feed room floor clear of spilled feed, thereby reducing feed for rodents and therefore feed for snakes in countries that have them.

- They pick up horse feed that falls out of the horse's mouth while eating. This reduces the chance of sand colic, as there is then no reason for the horse to sift around in the dirt/sand picking up bits of feed (and sand). Therefore, nothing is wasted and a potential problem is eliminated.

- Whatever they eat they turn into an extremely nutritious food for humans (eggs) and they provide fertilizer via their own manure.

- They help to desensitise horses to flapping things.

If you do keep chickens on your horse property, there are a few things that you need to be aware of:

- Chickens can carry salmonella, but then this can already be in the soil as well. Good hygiene will reduce the risks of salmonella poisoning in horses – for example, chicken manure must be kept away from loose horse feed including hay.

- Commercial chicken feed can be poisonous to horses; if you use it, keep it in a safe place where horses cannot get to it.

Chickens and horses have lived side by side for centuries.

- Chickens will eat dung beetles, so keep them away from fresh horse manure if you are trying to establish/encourage dung beetles on your land.

- Equine chemical parasitic worming pastes are poisonous to chickens. For several days after using such a product on your horses, collect the manure, bag it and take it to the refuse tip or at least prevent access to the manure for the chickens.

- Some horses are aggressive to chickens or are frightened of chickens, so be careful when first introducing them to each other.

- Most of the mites that chickens can carry do not affect horses, but a couple of them can so keep on top of them both for the sake of the chickens and the sake of the horses.

There are several ways that you can integrate chickens on to a horse property:

- You could build/buy a mobile 'chicken tractor'. For some ideas search 'chicken tractor' on the internet have a look at the images that come up. Basically, a chicken tractor is a chicken coup on wheels. This mobile chicken coup can be moved around the paddocks as the grazing animals are moved around the paddocks (as part of a rotational grazing management system). This method is being used by some farmers to control parasitic worms in their cattle, therefore reducing the need, cost and potential negative effects of chemical usage and to save on labour (because it spares them having to harrow the paddocks to spread the manure). If you have dung beetles, keep the chicken tractor out of the paddocks when they are working. (see the section *Dung beetles and other insects*).

- Instead of building a manure storage/composting area, you could build an area that has a moveable coup at one end and a 'yard' that has removable sides where you put manure that is collected from stables/surfaced holding yards and paddocks. The chickens will 'clean' the horse manure for you and they will also mix their own manure into it. After a period of time you can move the coup along to another area and you then have fantastic fertiliser to spread back on your land when a paddock is undergoing rest and recuperation (as part of a rotational grazing management system).

- Alternatively, you can grow vegetables straight into the raised area created by the chickens. You can then spread this 'used' compost on your land after using it to grow vegetables.

- Instead of a movable coup, you could have four chicken 'yards' that lead off from the coup in each direction. This way you do not need to be able to move the coup, you just use each chicken yard in succession.

—

There are some very useful links on the **Equiculture website page** about horses and chickens. See **www.equiculture.com.au/horses-and-chickens.html**

—

Using improved manure

The expensive nutrients in manure are wasted if manure is not utilised properly. Spreading manure (composted or otherwise) back on the land abides by what is known as 'the law of returns', in that what is taken away should be put back therefore, if possible, improved manure should be used on the pastures to replace some of the nutrients that grazing horses take out of the system. These products are very high quality, therefore you should plan where they will be used to best effect.

Spreading improved manure

Things to keep in mind are that:

- Manure should be applied no more than 2.5cm (1ins) thick on pastures and no more than 3 to 4 application per year.

- Spread it only during the growing season when it will be used by the plants and remember that good pasture can utilize compost much better than bare soil.

- Some localities have strict rules about spreading manure because of the risk of pollution, in particular, to the waterways. Check up on local regulations before spreading, as you may not be able to spread it at all or, if you can, you may only be able to spread certain amounts. Your local authority should know what the rules are.

- Manure should only be spread on land *after* horses have been rotated on to the next paddock (as part of a rotational grazing management system).

- Don't spread manure during very hot dry weather (or it will dry out and become dust) or in very wet weather (or it will wash away).

- If using manure on a patch of bare soil, it can be covered with less valuable organic matter, such as old but weed-free hay, to protect it from weather extremes so that it can condition and fertilise the soil without blowing or washing away.

Manure should only be spread on land after horses have been rotated on to the next paddock.

- Never spread any form of manure on water saturated ground or on land subject to flooding. Never spread manure near a waterway.

- If possible, spread it on high ground where it can be filtered and utilised by vegetation before it reaches any waterways. This way *more* of your land benefits from the soil conditioning and fertilising effects of the manure.

- No form of manure should be used as a surface in high traffic areas (such as for filling in low spots in a training yard or laneway) because manure holds a lot of water and, when it rains, this area will become waterlogged and full of bacteria, also, when it dries, the manure will become a fine powder that can be inhaled by people and animals.

- Manure can be spread by hand in a paddock with a shovel from a wheel barrow or trailer, or by a commercial manure spreader. It is now possible to buy smaller manure spreaders that have been designed for horse properties (most horse properties are smaller than most farms). These can usually be towed behind a ride on mower, four wheel bike, utility vehicle or small tractor.

- Avoid spreading fresh manure at the base of trees and bushes as it can kill them because it is too strong. Some trees cannot cope with the nutrients in improved manure either.

Spreading 'fresh' manure

Manure that is collected from stables/surfaced holding yards should ideally be improved in one of the ways outlined above (see the section *Improving manure*) before spreading it on land. However, some horse owners do spread it without composting it or otherwise first. **If you decide to simply spread it straight on your land, there are several points that you should take into consideration:**

- There may be regulations in your locality that deem this practice illegal. This is because if fresh manure gets into a waterway, it can cause severe pollution.

- Even if this practice is legal, be very careful about where you spread it. Follow the guidelines set out in the section *Using improved manure*.

- Never spread fresh manure in paddocks that horses are currently grazing, otherwise they will pick up large numbers of parasitic worms.

Preferably only ever spread fresh manure on land if it is also grazed by other animal species in order to reduce parasitic worms. In addition, spread it in a paddock that is scheduled to have a long period of rest, as part of a rotational grazing management system. Hot dry weather or very cold weather will also help to reduce the parasitic worms.

Managing surplus manure

Sometimes there is too much manure to spread on paddocks. This can occur when a large number of horses are kept in a relatively small area, such as an urban riding school for example. **There are various things you can do with surplus manure:**

- Manure can be put in special garbage containers from a waste disposal company and removed regularly (at a cost) from the property, but this should be the last resort as it is an expensive and unsustainable solution. Even on this kind of property, it may still be possible to compost the manure and then sell it, thereby turning a liability into an asset. Vermicomposting can work well in such a situation (see the section *Composting worms - vermicomposting*).

- Fresh or composted manure can usually be sold at the 'farm gate' because, in built up areas, there is usually a high demand for manure for gardens. Fresh or composted manure can be marketed to home gardeners, nurseries and crop farmers. When manure is to be used for gardening or growing crops, care should be taken to separate the manure from horses that have had chemical parasitic worming pastes used on them in the last few days, as this manure may kill earthworms. Thorough composting may break the chemicals down, but opinion varies as to whether it does.

- If it is not possible to market your manure in this situation, try to find a company that will either buy or remove your manure for free on a daily or weekly basis. In some areas there are 'manure banks' or similar composting schemes who allow you to contribute your manure, and later on 'draw out' compost. Contact your local authority to see if such a scheme operates in your area. If it does not already, enough interest from the community may mean that such a scheme gets underway.

—

There are some very useful links on the **Equiculture website page** that supports the subject of horses and manure **www.equiculture.com.au/horses-and manure.html**

—

Manure can be put in special garbage containers from a waste disposal company and removed regularly.

Grazing management

As a horse owner and land manager, you need to nurture and maintain your pasture in much the same way that a farmer would. Farmers require high energy (high sugar) plants for maximum productivity, conversely most horse owners/managers require low energy (low sugar) plants. However, the principles for managing pasture plants is the same in many aspects, whether you want to grow high energy or low energy plants.

The grazing management systems outlined in the following sections are all variations on the same theme of restricting horses to one part of the land (even though this may be a surfaced holding yard), while the other parts get to rest, recuperate and grow more pasture.

A basic rule of thumb is that *no more than 30%* of the land should be in use at any one time (see the section *Pasture plants in their natural environment*). It is common on a horse property to have *all* of the land in use *all* of the time, either by letting the horses graze the whole property as a herd (e.g. not restricting them to any one paddock) or by separating horses into individual ('private') paddocks which are in constant use. This is called set-stocking and it is not recommended as a good way to manage land (see the section *Set-stocking*).

Even if pasture is irrigated and fertilised it will not be able to cope with continuous pressure from the hooves of horses and their ability to eat plants right to the ground - it needs time to recover.

The various grazing management systems outlined below (with the exception of set-stocking) should be used in conjunction with one another for best results. Aim to be flexible and be prepared to change what you do to suit the current situation. For example, seasonal changes and uncharacteristic weather for the season (such as a huge downpour) will call for changes in the day-to-day routine.

A surfaced holding yard is required so that the horses can be safely confined when necessary (see the section *The importance of surfaced holding yards*). Failing that, you will need to use stables if the pasture needs a reduction in grazing pressure. Grazing time can then be increased when pasture is available and decreased when it is not. Supplementary feed such as hay and possibly concentrates should be used to make up the shortfall (see Appendix: *Feeding confined horses*). It is far better to confine your horses some of the time, if necessary so that the time they spend in the paddocks is 'quality time' (moving and grazing grass), rather than have your horses standing around all day in bare, dusty, muddy, weedy paddocks (making them even more bare, dusty, muddy and weedy).

By utilising good grazing management your horses will benefit by having more grass available to them and overall increased turnout times. Think in terms of quality grazing time. Land degradation will be reduced due to maximising the potential of the land without depleting it.

A surfaced holding yard is required so that the horses can be safely confined when necessary.

Stocking rates

The actual amount of animals a particular area of land can carry varies enormously, depending on many factors such as soil type and fertility, type of pasture plants, annual rainfall and climate etc. Because land varies so much, land managers use the term 'stocking rate' (or 'livestock unit' in some countries) as a guide when describing land. The stocking rate is the number of animals a given piece of land can support productively and without land degradation. It is a guide for land managers that helps them to gauge how many animals a particular piece of land will support without having to overly rely on supplementary feed. Supplementary feed is expensive and can make farming non-viable. Stocking rates are important for farmers who have to balance out the time/energy/money spent on land and animal management with its returns. So for example, if a farmer was considering buying an area of land, finding out the stocking rate for that land would be an important consideration.

In the case of horses and their management, the stocking rate can be a useful guide, but there are many other things to take into consideration.

For most horse owners, financial return is not the primary reason for keeping horses and they are usually prepared to invest time and money without an economic return from their animals. They keep horses for emotional/pleasure driven reasons rather than economic reasons. Horses are kept as pets and as competition animals and as everything in between – but rarely for meat production. Good horse owners can and will 'take the rough with the smooth' because it is not usually their livelihood that is directly at stake.

Horse owners tend to keep the same animals for long periods of time, whereas farmers usually turn animals over more quickly and can reduce numbers at certain times of the year, or if certain conditions such as drought are forecast. Horse owners usually expect to have to supplementary feed their animals – at least from time to time.

Horses can and often are kept using a variety of management systems which vary in intensity from 100% at pasture to 100% confinement, although this is certainly not recommended, with many variations in between. Therefore, the stocking rate is just a guide which becomes more important the more pasture is relied on as a feed source.

The stocking rate is calculated differently in different countries. The stocking rate of an area can usually be obtained from your Agriculture office (or a land/soil conservation group). For more information on stocking rates it is suggested that you look up 'Livestock Unit' on Wikipedia.

Once the stocking rate is known, you can work out what your land *should* be capable of holding on a yearly basis, assuming that your land is managed well and there is not a drought etc. Also, if an area is calculated as being able to support three horses full time, it can support six horses for half of the time, although that depends on the time of year and so on. Unless your land is already producing pasture at a very high level, the stocking rate can be improved with good management.

However, you must keep in mind that every day, every month and every year is different. You have to be flexible if your horses are not to get too fat or too thin and if your land is not to become degraded. This is where grazing management systems come in.

When calculating your stocking rate, remember that between one and two acres will usually be taken up with buildings and infrastructure including the driveway, any dams, stables, yards, an arena etc. therefore the area left over is what you will have available for grazing. Bear in mind that some areas will only be able to be grazed at certain times of the year e.g. some areas may be wet/waterlogged for part of each year and fragile areas (such as sandy soil) will need more protection when the weather is hot and dry.

The amount of animals that you can keep on your land may also depend on local regulations, so you need to check with your local authority. Sometimes a local authority has regulations about how many horses can be kept on a given area, odour pollution, manure management, flies and drainage. As time goes on, it is likely that more and more local authorities will adopt policies for horse control due to the increasing number of horses being kept in the 'urban fringe'. By managing your land well and involving other like-minded people who live in your area so that they also do the same, you may be able to stave off the time when the local authority dictates what you can and cannot do.

In the case of horses and their management, the stocking rate can be a useful guide, but there are many other things to take into consideration.

Grazing management systems

Set-stocking

Set-stocking means that the grazing animals on a given area of land are allowed access to all of it, all of the time. This means that all of the land in question is in continuous use.

On a typical horse property this usually occurs in one of three ways:

- The horses are separated into individual 'private' paddocks and each horse has continuous access to that paddock on a daily basis - either 24/7 or, for example, confined at night in a stable or surfaced holding yard and out in the paddock by day.

- The horses are kept in a herd but have continuous access to the whole of the land on the property - e.g. either there are no internal fences or the internal property gates are always open.

- The horses are segregated into various separate groups, e.g. mares and geldings and, in particular, equines that have 'issues' with their weight, resulting in all of the available grazing areas being used at the same time.

The first scenario commonly occurs on commercial horse livery yards where owners actually want their horse/s to be kept separate from other horses, thinking that this is the safest option. Whether they are safer or not is debatable, because an increase in fence injuries is just one consequence of separating horses. Another reason for doing this is to make it easier (and safer) for individual horse owners to access their horse. Private horse property owners often also separate horses in this way, for the same reasons as above. A factor that can lead them to doing this is because they may have been exposed to this management regime if they kept their horse at livery before buying their own horse property.

The second scenario usually occurs when a land owner/manager does not understand the benefits of allowing pasture time to rest and recuperate. Also, they may feel that it is more 'natural' to allow their animals to have access to a larger area (see the section *Grazing management* for an explanation of why this does not work in the domestic situation).

The third scenario, because it involves several groups of horses, means that it is even more difficult to rest any of the land. In addition, horse owners may deliberately over-graze their land, believing that this is the best way of managing overweight horses (by reducing their pasture intake), **but this does not necessarily work for various reasons:**

- Equines are designed to, and are good at, eating short grass. Their lips, teeth, grazing behaviour enable them to consume a lot of energy when grazing short pasture plants.

- We now know that very short grass is stressed grass and stressed grass is usually very high in sugar/starch. Set-stocking does not work well on a horse property - unless you have a *lot* of land, and even then you need other grazing animals in order for it to work (see the section *Cross grazing*).

The practice of set-stocking is why so many horse properties appear badly managed. This practise leads to unhealthy land and unhealthy horses as the land becomes degraded over time due to never being allowed to rest and recuperate. A further problem with using this system is that manure must be picked up because harrowing a paddock that contains horses will increase their parasitic worm burden. All in all, set-stocking is to be avoided as a management practise on a horse property.

Set-stocking means that the grazing animals on a given area of land are allowed access to all of it, all of the time. The picture below shows how an area of land that is set stocked with horses in 'private' paddocks, starts out and ends up.

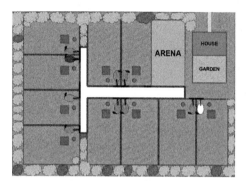 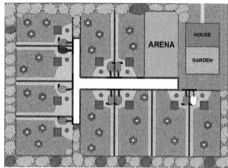

Rotational grazing

Rotational grazing means that the horses are kept in herds and are rotated around the land so that pasture gets periods of rest in between periods of use. It is the primary system that you should use to manage your land. This grazing management system means that pasture is subjected to a similar type of grazing pressure that it would have in a natural situation (see the section **Pasture plants in their natural environment**). Therefore it responds more naturally .

Paddock rotation allows grass species to recover where they would otherwise die out if submitted to constant grazing pressure. Horses tend to eat only what they like (find palatable) and leave the other species. This results in short, stressed grass with the possibility of certain species, including weeds, taking over the pasture if it is poorly managed.

Rotational grazing greatly improves biodiversity (see the section **The importance of biodiversity**), pasture growth and parasite control and reduces land degradation. This method of management also helps to prevent the under/over-grazing pattern present in so many horse paddocks.

Horses should be allowed to graze a designated paddock when the plants have reached an average height of about 15cm (6ins) to 20cm (8ins). They should then be moved to the next paddock when they have grazed the paddock to an average height of about 5cm (2ins) and 8cm (3ins).

Initially you may have to use a measure, but after a while you will be able to assess the length of the grass by eye.

When the animals are moved on, the now empty paddock can be 'tidied up' if needs be (see the section **Pasture maintenance**) and then rested and allowed to regrow. When the pasture has reached the appropriate length again, the horses can once more graze the paddock.

If the situation occurs where none of the paddocks are recovered enough for grazing, the horses should be confined to surfaced holding yards until they are.

If re-growth consistently fails to keep up with grazing demand then limited access to pasture (see the section **Limited grazing**) should be used in conjunction with rotational grazing.

If you have no alternative, it is better to graze a rested pasture earlier, before it has reached the optimum height, than to allow horses to overgraze a pasture that is currently in use too short. Overgrazing should be avoided at all costs.

The length of time that it takes the paddock to recover to an acceptable grazing length depends on factors such as the time of the year, the amount of rainfall and the growth period of the particular pasture species. Different climates tend to receive their rainfall at different times of the year, with temperate regions having wet winters and sub-tropical or tropical regions having wet summers.

The amount of time that a pasture can be grazed *without damage* will vary throughout the year, and likewise from year to year, depending on climatic changes (such as drought and excessively wet years). Pastures need a certain amount of warmth and water in order to grow so, in some areas, there can be prolonged periods of little or no growth even in a 'normal' year.

It does not work to graze/rest each paddock for a *set* amount of time. For example to divide your land into six paddocks and graze each of them for two months before moving the horses on would be too rigid a management system. There will be times when two months is too long. Instead, you need to gauge each paddock by eye and decide when it is time to move the horses on.

An important factor is that the empty paddock receives a couple of bouts of rainfall to encourage growth, followed by sunshine (or frost) as this will reduce parasitic worms in the pasture. Rain causes the worm eggs passed in the horse's dung to hatch out into larvae; they then climb up a blade of grass and wait for a horse to eat the grass. If horses are not currently grazing the paddock, some of the larvae die due to the sun drying them out. Frost can also kill them during colder times of the year.

Rotational grazing means that the horses are kept in herds and are rotated around the land so that pasture gets periods of rest in between periods of use.

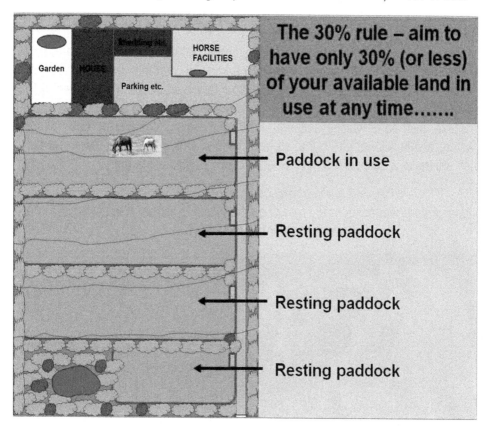

Limited grazing

On most horse properties, there is a limit to the amount of available pasture and a combination of grazing and confinement will need to be used. The term 'limited grazing' means that you allow the horses to graze a designated paddock for only part of each day, then confine them to a yard with low energy hay for the rest of the time. So, you might keep the horses off the pasture either overnight or through the day for example.

This strategy is carried out on a horse property for two main reasons. 1. To reduce grazing pressure so that land does not become overgrazed and 2. To reduce the pasture intake of horses (although this does not always work as intended - see below). In some areas, horse owners also confine horses overnight for their safety e.g. because of the presence of wild/feral animals etc.

Removing horses from a pasture for a period each day vastly reduces the amount of time they spend sleeping or loafing on the pasture, thereby reducing land degradation and allowing more flexibility in the amount of horses that can be kept on a particular area of land. When horses are removed from pasture for a period of time and are then returned, they tend to get straight down to grazing.

Pasture quality and quantity varies markedly throughout the year. This will influence the need for provision of supplementary feed to pastured horses. It is a good idea to use hay, fed in yards, to reduce the grazing pressure whenever possible. Instead of waiting until your land is struggling to cope, feed hay and the pasture plants will not get to that stage. In addition, due to having more leaf area, they will make a quicker recovery when growth conditions improve.

The term 'limited grazing' means that you allow the horses to graze a designated paddock for only part of each day, then confine them to a yard with low energy hay for the rest of the time.

Keep in mind though that a horse will concentrate their grazing into the period that they are turned out. So, restricting a horse by removing them from pasture and then turning them out does *not* necessarily reduce their intake, unless the turnout time is very short. For example, if you confine a horse for eight hours and turn them out for 16 hours, the horse will concentrate their grazing bouts into this 16 hour period. Therefore, doing this could result in them gorging themselves when turned out. In fact, a recent study showed that ponies that were confined for 21 hours (with 'ad-lib' hay) and turned out for just three hours per day were able to eat *three quarters* of their daily intake in that three hour period! For most horses, once

128

they accept that fibre is available 24/7 in the form of hay or grass, their overeating behaviour tends to stabilise. However, one theory suggests that we could be creating eating disorders in our horses by using restrictive grazing practices, so it is important to monitor their intake, especially when changing grazing practices.

All horses, even those with weight problems, must be given forage when confined. The forage type should reflect the weight condition of the horse in question – for example a horse that does not need to gain weight can be given low energy hay, whereas a horse that needs more energy can be fed higher energy hay and maybe even concentrates. Horses should not have long periods without forage, as the gut is not designed to cope with being empty. See Appendix: *Feeding confined horses*.

Another alternative is to let the horses graze for two shorter periods per day, rather than one long one, but this means that the horses have to be led in and out two times a day rather than one. **The Equicentral System** allows the horses to come and go as they wish, which means that they *voluntarily* remove their grazing pressure and, if you wish, the horses can *also* be confined for part of each day (e.g. overnight) for the reasons outlined above (see Appendix: *The Equicentral System*).

As pasture is improved, time spent grazing can be increased and reliance on supplementary feed reduced. It is far better to confine horses some of the time, so that the time they spend in the paddocks is 'quality grazing time' (walking and eating pasture), rather than standing around all day in bare, dusty or muddy paddocks (making them even more bare and dusty or muddy).

Although horses eat for about 12-16 hours on average (more in adverse conditions), 8–12 hours is spent on doing other behaviours such as sleeping, drinking and social behaviour. If we can encourage horses to do these behaviours off the land, we are reducing the grazing pressure by about 40% (remember grazing pressure is not just the damage done by their teeth, but also by their four hooves). Using **The Equicentral System** the horses voluntarily remove themselves from the pasture when not grazing (see Appendix: *The Equicentral System*).

This grazing management system should not be used *instead* of paddock rotation, paddocks still need periods of total rest, but can be used to *compliment* paddock rotation. This grazing management system can also be used at the same time as other grazing management systems such as strip grazing and cross grazing.

Limiting grazing is an excellent strategy for reducing land degradation *and* for making your available pasture last as long as possible. It may not be as effective at controlling your horse's intake, although if necessary, you can reduce intake to some extent by reducing the turn out period significantly.

Strip grazing

This strategy involves using portable electric fencing to reduce the area of the paddock currently available to the horses. It is commonly used with cows in the dairy industry to make the available pasture last as long as possible. Many horse owners also use this grazing management strategy. Horses are very selective grazers and, if they have access to a large area for too long, they will overgraze some areas and under graze others. Eventually, the land will develop that characteristic 'horse-sick' appearance (see the section *The 'dunging behaviour' of horses*). By having their access restricted to a smaller area, the horses will graze more evenly on the available pasture and, at the same time, the fenced off area is allowed to be in the rest and recuperation phase for longer, resulting in more pasture growth. Strip grazing is actually rotational grazing to an even greater degree.

Horse owners are sometimes concerned that reducing the paddock size will reduce the horse's movement but, in fact, as pasture management practices, including strip grazing, encourages more biodiversity by allowing different plants to survive and thrive, this encourages movement and horses move more on biodiverse pasture than they do on a 'monoculture' of just one type of pasture plant. When grazing a range of plants, horses keep moving, looking for different plants. When they graze a monoculture, they do not move as far once they have worked out that all of the pasture is the same; they tend to spend most of the time close to the gateway.

You can move the fence daily, every few days or weekly, depending on the size of the area, the current growing conditions, the amount of horses etc. Take care that you do not allow the area that is currently being grazed to be overgrazed. You still need to adhere to the same plant lengths as for rotational grazing (see the section *Rotational grazing*).

Where you put the temporary fence will depend on the shape of the paddock and where the water is. A square paddock with the water and gateway in or near one corner can be strip grazed by 'fanning' the fence out from the corner (make sure you round off any acute corners with electric tape). A long narrow paddock can be strip grazed by starting at one end and moving the fence gradually across the paddock, so that the available grazing area gets larger; you will still be doing a good job because some of the pasture will still spend longer in the rest and recuperation phase. You do not have to be too pedantic about moving the fence a measured amount every day as farmers do in the dairy industry for example (unless you want to), because simply subdividing a paddock will result in more even grazing and less wastage than if you were *not* to do it.

You can utilise **block grazing** and **temporary laneways** for more benefits if you wish.

A square paddock with the water and gateway in or near one corner (to the right of the waterway in this illustration) can be strip grazed by 'fanning' the fence out from the corner.

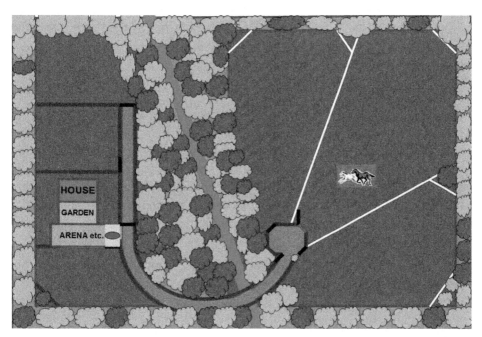

Block grazing

If the paddock is going to take several weeks to graze using strip grazing (due to the paddock being large), a second electric fence can be added behind the horses so that they cannot go back over the area that they have already grazed. This area can then be maintained (harrowed, mowed etc.) and be in the rest and recuperation stage. This strategy is only viable if the shape of the paddock lends itself to it. Fanning a fence out from a gateway can be a good option as there can then be one fence preventing the horse from moving to the fresh pasture and one fence preventing the horses from grazing back over the already grazed area. If the paddock is long and narrow, block grazing is more difficult and a temporary laneway might be a better option.

Temporary laneways

See Appendix: **Temporary laneways.**

Cross grazing

Cross grazing (sometimes called 'mixed grazing') means that different types of grazing animals, as opposed to just horses, are used to graze an area of land. Different types of animals grazing land means that the plants are being grazed in a more natural way, and respond accordingly (see the section *Pasture plants in their natural environment*). In nature, many different species of animals occupy the same areas of land and therefore by utilising cross grazing, you will come closer to replicating a more natural management system and therefore you, your animals and the wider environment will reap the benefits.

Cross grazing has two main advantages:

- Different animal species tend to complement each other in their grazing behaviours by eating different plants. Sheep and goats, in particular, will eat woodier plants that are often left behind by horses, and can also safely eat some weeds that can be harmful to horses. As these other animals eat the parts of the pasture that the horses tend to ignore (including the 'roughs'), having more animals does not always mean more total grazing pressure as it would if you simply added more horses. Cross grazing results in more 'even grazing' of a pasture e.g. reduces 'roughs' and 'lawns'.

- Different animal species will eat around the dung of other species, but not that of their own. This is thought to be a parasite prevention strategy because most parasitic worms are host specific, they can only complete their life cycle in one species of animal. Therefore, grazing animals instinctively avoid the dung piles of their own species, but not those of other species. If a parasite is picked up by the 'wrong' animal, the parasite dies in their digestive system meaning that cross grazing reduces parasitic worms on your land for all of the different grazing animal species.

Things to keep in mind are that...

Some of the disadvantages of utilising cross grazing are that there will be more mouths to feed when the pasture runs out, and that there are extra expenses involved for the additional animals such as foot care and possibly shearing. You will need to learn about the needs of these different animals before taking them on – for example, you will need to have several cows or sheep, not just one, because these animals need the companionship of their own species – just as horses do. Fencing can be an issue, especially if smaller animals are to be used. Horses can and sometimes do bully other species, so be aware.

If your neighbours have different animal species to yourself, this can be a good solution as you may be able to 'borrow' them occasionally and reap the benefits, without having the long term responsibilities.

You can raise your own meat if so desired, allowing you to remove the grazing pressure at strategic times (e.g. before the pasture becomes overgrazed) as the meat animals can be removed for processing. This also brings the advantage that you can make sure the meat you are eating has been well cared for.

Keep in mind, horses are susceptible to Liver Fluke (a parasite affecting a range of livestock and other species). Final hosts in which it can develop to sexual maturity include livestock such as sheep, cattle, horses, pigs, goats, alpacas and deer. Liver Fluke occurs where the intermediate host (lymnaeid snails) are present. These snails are found where there is slow-moving water, swamps etc., and they can survive in mud when water flow temporarily stops. However, the snail is not necessarily present in all such areas. Speak to a local vet if you are concerned.

Cross grazing (sometimes called 'mixed grazing') means that different types of grazing animals, as opposed to just horses, are used to graze an area of land. Take note in this picture - the land is also being strip grazed, the sheep having access to the land to the left, before the horses by way of a 'sheep-friendly fence'. The sheep can also use this area to get away from the horses if the horses start to bully the sheep.

The order of grazing:

When utilising cross grazing, the other animals can be grazed in a paddock before the horses, at the same time or after.

There are various advantages and disadvantages to keep in mind:

- It is commonly believed that putting other animals in a paddock *before* horses will reduce the pasture intake of the horses when they are allowed in to graze the paddock, because the other animals will have eaten most of the grass. This idea comes from the inaccurate belief that horses eat less when grazing on short grass, whilst in actual fact, they have the ideal equipment to eat short grass. It is now known that shorter grass is higher in sugar *per mouthful* (see the section **The importance of healthy pasture plants**). Sheep in particular will shorten and 'sweeten' the plants so if you have horses that have 'weight issues', grazing other animals on a pasture *before* the horses might not be a good idea.

- Grazing the other animals in a paddock *at the same time* as horses is fine as long as they have plenty of room to keep out of each other's way. Horses are usually dominant over other grazing animal species, in particular because they still have their weapons intact (e.g. they are often shod, they can kick and bite) whereas cattle and sheep usually have had their horns removed and they are not much good at kicking. These animals are also not able to run as fast as horses and so cannot get out of their way! Horses are more active than most other species of grazing animals and will sometimes actually chase other animals that are in the same paddock. Generally speaking, if a paddock does not have enough space for the different animal species to keep well out of each other's way, then graze them one after the other rather than at the same time.

- Grazing other animal species *after* the horses will tidy up the pasture and reduce or eliminate the need for you to mow. The other animals should eat the long grass that the horses have left behind; cows in particular should eat the 'roughs'. If the 'roughs' are very well established e.g. they have not been mowed or grazed by other animals for some time previously, the plants in them will be older and rank, so the cows may ignore them. You may need to 'freshen' them by mowing them once or twice before the cows will graze them. Once you are regularly putting cows in after the horses this should not be an issue.

Which other animals to use?

When deciding which type of animals to use for cross grazing there are a few factors to consider. The most common animals used for cross grazing with horses are cows and sheep, however many other grazing animals can also be used.

Note: Donkeys do not count as a potential animal for cross grazing due to them also being equines.

Sheep in particular will shorten and 'sweeten' the plants.

Cows

Cows are the most usual choice for cross grazing:

- They have similar fencing requirements to horses. They do tend to put more pressure on fences than horses, but well-constructed electric fences work as well for cows as they do for horses. You may need two different heights of electric. For example, if you have a post and rail fence you will need one electric element high up for the horses and one lower down for the cows.

- Horses graze with their front teeth, biting the grass off close to the ground; cattle, on the other hand, tear the grass off with their tongues and require longer grass. This means that cattle tend to eat what horses leave. On a small horse property, it can work well to either put cows in the paddock after the horses have been rotated to the next paddock to 'clean up' after them *or* to graze the cows with the horse herd, moving around with them. Keep in mind though, even adult cows can be bullied by horses and calves in particular can be injured or killed by a horse. Some horses, such as trained cutting horses can be particularly 'bossy' with cows.

- Putting cows in the paddock *before* the horses in order to reduce the amount of feed (in the case of fat horses for example) does not necessarily work. Grass has to be *extremely* short before it starts to reduce the intake of horses. When it is short it is also relatively higher (higher per mouthful) in sugar and starch.

- Cows will reduce the clover in a field if grazed at the correct time.

Cows are the most usual choice for cross grazing, this combination is often used in conservation grazing projects.

Sheep and goats

Sheep and goats are the next most usual choice for cross grazing:

- They will tackle most weeds found in horse paddocks and they also tend to be *less susceptible* to the poisonous effects of certain weeds.

- Goats are particularly good at eating plants such as Gorse and Blackberry and have the advantage over sheep in that most goat breeds do not require shearing (if you choose the correct breeds), however some breeds of sheep do not require shearing either (e.g. Wiltshire Horn).

- Larger breeds of goats are easier to keep in the paddock than smaller breeds. Wethers (castrated males) tend to be more docile.

- Sheep and particularly goats, if handled young enough, make good pets.

- Sheep will tend to keep a pasture short and level, meaning the grasses will be sweeter and as far as horses are concerned, nicer. Just be aware that horses prone to laminitis may be at more risk on such pastures.

- Electric fencing does not usually control these animals and they may need netting to keep them in.

- Beware of foot rot in prolonged wet weather.

- Goats need shelter from rain, as they originate from mainly arid environments, whereas sheep are hardier in bad weather.

- Being small and cloven hooved animals, sheep and goats do not cause as much damage to the land as larger, heavier animals such as horses and cows.
- They are generally inexpensive to acquire.

Sheep and goats are the next most usual choice for cross grazing.

Lamas and alpacas

Lamas and alpacas are a less common choice for cross grazing but have many advantages:

- They cause less damage to the land with their feet than hoofed animals; they have pads rather than hoofs.
- They are generally placid and make good pets.
- They have the unusual benefit of being able to protect other animals such as goats, sheep, chickens and possibly foals from predators (such as wild/feral dogs), and are being used for this purpose on some farms. They also tend to be reasonably confident around horses.
- They will eat a variety of plants including woody weeds.

Things to keep in mind are that:

- Even though they require shearing, you may be able to get a local breeder to do it for free in return for the fleece.
- They are generally more expensive to buy then sheep or goats. Wethers (castrated males) are the cheapest, with breeding animals being much more expensive.

- They need shelter from the elements – especially rain - as they originate from arid environments.

Poultry

- Chickens can be good for pasture although there is a limit to how much grass and weeds they can eat. They work well in conjunction with other cross grazing animals as they scatter manure whilst searching for seeds and parasite eggs. Geese and ducks also eat grass. Muscovy ducks eats pest insects as well. Geese have the added advantage of acting as an alarm service; they will let you know when you have visitors. Birds must be supplied with their own dam or pond and prevented from using the dams that supply the other animals' drinking water, as their droppings will cause algae. Some horses dislike chickens, geese and ducks to the extent that they will chase and stomp them so be careful.

Large birds

- Emus and ostriches can be also be used as cross grazing animals. They provide several benefits; their feet cause less damage than those of hoofed animals, they produce good manure, they scatter horse manure, and also eat some harmful pasture insects and weeds, however horses must be habituated to these large scary birds. Emus and ostriches can usually be kept in with horse fencing. You will need to check your local authority for regulations etc.

This picture was taken in Scotland! Even though the Donkeys do not provide cross grazing benefits, the Emu certainly would.

Wild animals

- Wherever you live wild, native animals will also be cross grazing your pasture. Always make sure your fencing allows native animals to pass through safely.

Implementing cross grazing

Borrowing animals for cross grazing may be a better option than actually owning more animals; this way the extra animals need only be on the land as and when you deem it necessary. You could either share some animals with a like-minded friend or neighbour, or a local farmer may be happy to occasionally put some animals in your paddocks.

In summary, cross grazing should only be considered when a property provides abundant feed for most of the year, rather than a property that struggles to provide enough feed for its present occupants. If a property is regularly producing *too much* grass for the current occupants, then cross grazing is a better strategy for land management than acquiring yet more horses. In order for this grazing management system to work well, the land needs to be producing grass at its optimum level and, for this reason, it is a strategy that can be employed at a later date when the land has been improved. If cross grazing is to be employed on a horse property, it is best to plan ahead with regards to fencing, so that expensive alterations do not have to be made at a later date.

Pasture maintenance

Pasture maintenance is the name for routine procedures that may need to be carried out in order to maintain or improve the current condition of the pasture. Typically it includes mowing the remaining plants (after grazing), harrowing manure (or removing it – see the section *Picking up manure in paddocks*) and weed control (see the section *Weeds*). You may not need to do all or even any of these; it depends on your individual situation. This section goes through some of the pros and cons and gives reasons for why you may or may not need to carry out these jobs.

Mowing

There are various terms used for the practice of cutting pasture plants, namely mowing, slashing and topping. Slashing and topping are the same thing, so for this section we will call it topping. The difference between topping and mowing is that topping simply cuts the plants once, leaving long stems lying on the ground, whereas mowing chops the plants (mulches them) into small pieces. The latter is usually preferable, as the cut plant then decomposes sooner. The smaller clippings fall between the stems that are left standing and do not impede their growth (as tall 'topped' grass can do until it decomposes), protecting any bare soil from the sun/wind and adding organic matter to the soil as they decompose. A tractor can be fitted with a topping or mowing implement (mowing implements usually cost more), whereas domestic lawn mowers and 'ride on' mowers mow.

Reasons to mow:

- Mowing 'grooms' the paddock, leaving it looking tidy.

- Mowing encourages pasture plants to thicken and improves soil coverage. This results in denser coverage, which means that each plant will receive relatively less sunshine once it regrows because, as the plants grow, the individual plants will be closer together and will shade each other out. Therefore, the plants will be relatively lower in sugar/starch (NSC) and higher proportionately in fibre.

- Mowing allows sunlight to reach the base of the plants, giving shorter plants an opportunity to grow (as opposed to them being shaded out by taller plants).

- Organic matter (from the cut plants) ends up on the ground and decomposes to improve the soil, therefore mowing a pasture during a period of rapid growth can also be advantageous in terms of creating more organic matter for soil if this is needed.

- At certain times of the year, you may have more pasture than your animals can handle, but cutting the plants does them good - it copies what happens to them in the natural grazing situation. Mowing puts plants back into the growing phase because once they go to seed, they stop growing, which creates more organic matter (both above and below ground). The pasture can be cut to any height between 5cm (2ins) and 10cm (4ins) in order to have a beneficial effect.

- Mowing also helps to control certain upright weeds (but not prostrate weeds), because many weeds do not thrive when cut back. Cutting a *grass* plant mimics the grazing of an animal and stimulates the plant to start growing again. Therefore, timely mowing favours grass plants and gives them a competitive advantage over many weeds (see the section *Control of weeds*).

- On a small horse property, a ride on mower is generally all that is required to tidy up a paddock, because the horses will have eaten the bulk of the plants before they are moved on to the next paddock (as part of a rotational grazing management system). Ideally, the plants should be cut at approximately 5cm (2.5ins) or higher. If a ride on mower is used, it must be set as high as possible because, unlike lawn type grasses, most pasture grasses do not cope with being cut too short.

- Cutting pasture plants with a machine helps to remove old, dry grass and encourages the growth of new fresh leaves, resulting in more uniform regrowth. Mowing 'tidies up' the pasture and results in all of the plants having to regrow from the same starting height.

- Cutting pasture plants *just before* they set seed will encourage them to keep growing by keeping them in the elongation stage.

- Cutting any grass that is left behind *after* a grazing period is a way of artificially replicating what happens to plants in the wild situation. In the wild, herds of animals graze, trample and drop manure, but keep moving forward. The area then gets some rest and recuperation time before the next herd of animals, which are usually a different species, come along.

- Grass that is actively growing is actually 'safer' than grass that is too short or that has gone to seed. When using the word 'safer' here, we are speaking in terms of sugar and starch per mouthful, as the plant is using up its reserves of sugar and starch to grow. However, most horses will still need to have their intake monitored.

There are many reasons to mow a pasture including to produce a denser sward, to tidy up a pasture or to keep a pasture in the growth stage.

Why you might not need to mow:

- If you are cross grazing (see the section **Cross grazing**), you may not need to mow because the other animal species will have grazed the paddock more evenly than if it had just been grazed by horses.

- If you only have a very small area of pasture and you are picking up manure, the 'roughs' may not become established and you *may* not need to mow.

How/when to mow:

- Your pasture should be mowed immediately after the horses have been moved on to the next paddock (as part of a rotational grazing management system). The remaining pasture should be cut to a height of about 5cm (2ins), level with the rest of the paddock, presuming that the horses were moved on when the paddock reached this stage. Even when you are using other animals, it may be beneficial to mow the paddocks periodically; it just will not be required as often.

Things to keep in mind are that:

- Take care if you are cutting a paddock that horses are currently grazing; if the plants are topped rather than mowed, the horses will probably leave the cut plants, preferring the shorter grass left behind. However, be aware that this pasture plant stubble may be relatively high in sugar/starch until it has a chance to get growing again. If the horses do choose to eat the long pieces of cut plants, they still have to chew them.

- If the paddock is mowed instead, the very small pieces left behind may be eaten by the horses. This can be a problem because when horses eat grass clippings they are at risk from gut rupture or colic (see the section *A word about lawn mower clippings*). So, it is best to give the paddock at least few days rest after mowing to allow time for the clippings to either dry out, wash into the soil or decompose to a point that the horses will not eat them.

- Generally speaking, it is best to only mow a paddock when horses have been moved on to the next paddock (as part of a rotational grazing management system).

Take care if you are cutting a paddock that horses are currently grazing.

Harrowing

Harrowing is a manure management strategy. It involves dragging an implement around land that has been grazed (and has manure piles in place) to break up the manure piles and spread them around the pasture. Harrowing is a controversial subject, with some people being wholeheartedly for it, and some being wholeheartedly against it. In reality, both sides of the argument are correct – in certain situations. This section will give you more information so that you can make up your own mind and decide whether it would suit your own management system.

Remember: if you do not do something with the 'roughs' in a paddock, these areas become unavailable as grazing areas and therefore tend to become larger over time. For example, a paddock that is five acres may end up having only three acres or less available for grazing. This compounds the nutrients in these areas, with the 'lawns' becoming increasingly depleted as horses take nutrients from them (by grazing them) and deposit them in the 'roughs' (in their manure).

Reasons to harrow:

- Spreading manure around a paddock rather than allowing it to stay in the areas that horses deposit it (the roughs) results in more uniform grazing in the future (see the section ***The 'dunging behaviour' of horses***). The whole paddock benefits from the nutrients and organic matter in the manure. The nutrients in uncomposted manure are not as available to plants as the nutrients in composted manure (see the section ***Composting manure***), but it still means that they are better utilised than if they were left in the roughs.

Harrowing is a manure management strategy. It involves dragging an implement around land that has been grazed.

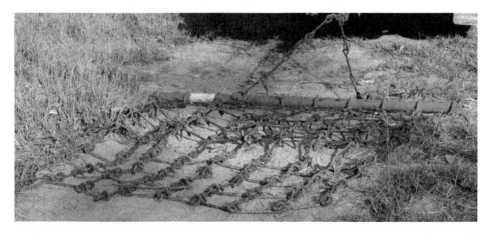

- Some people believe that harrowing is not a natural process, but in fact it is. In the natural situation, large herds of animals move across grassland and kick manure around with their hooves; harrows simply copy this process. Other species such as birds or beetles also spread the manure.

- Harrowing can help to kill some of the parasitic worms in the paddock. Breaking up manure dries it out and the eggs and larvae of parasitic worms need lots of moisture to survive, so when manure is drier *some* of the worms are killed. Strong sunshine or frost also kills some of the worms.

- Harrowing can also pull out any dead grass plants.

- Any subsequent rainfalls cause surviving parasite worm eggs to hatch. The larvae then climb to the top of grass stalks and wait to be eaten by a horse. If the paddock is being rested at this time (as it should be after harrowing as part of a rotational grazing management system), this causes *some* of the worm larvae to die as they do not survive in the larval stage outside of a horse for as long as they do as eggs.

- Some people believe that it causes horses to reject all of the paddock rather than just the 'roughs'. It is 'intact' piles of manure that horses avoid eating near so, as long as manure is broken up, spread and then given time to wash into the soil, harrowing results in more even grazing in the future.

- Harrowing takes much less time than the regular picking up of manure.

Reasons not to harrow:

- You may not have the budget to acquire them or the vehicle to tow them.

- If your local climate is mild for much of the year, it may not get hot, dry or cold enough to kill the parasitic worms before the horses are rotated back into the paddock in question. However the pasture itself benefits from spreading the nutrients.

- Some people enjoy picking up manure as they find it good exercise and therapeutic!

- If your grazing area is very small, harrowing might not be feasible. For example, if you rent a small paddock as is often the situation with livery, then you may not be able to rest areas of the paddock long enough to keep your horse/s safe from parasitic worms. In this case you will need to pick up manure from the land, see the section *Picking up manure in paddocks.*

- When **dung beetles**, the worlds class experts at dung management, are working.

How to harrow:

- Pasture harrows can be purchased in all shapes and sizes, they can also be home made. Homemade harrows can be constructed from many things such as from the springs of an old bed base, chain link fencing weighed down by tyres etc.

- The aim is to ensure that each and every pile of manure is broken up and scattered. If your land is not level, you need harrows that are flexible and can get into all the contours of your land. Commercial harrows usually have spikes that scratch the surface as spreading manure. This can pull out old dead plants and create grooves for new seeds to become established. But it is the manure spreading function that is the most important.

The aim is to ensure that each and every pile of manure is broken up and scattered.

- Harrows can be pulled by a variety of vehicles ranging from a tractor to a four wheel drive, an ordinary car (in the right conditions) to a four wheel bike (for very lightweight harrows).

- Drive reasonably slowly otherwise the harrows will 'jump' over some of the piles.

- Make sure you store harrows in a safe area; harrows are notorious for being left in paddocks and then forgotten about because the grass grows through them and then horses can inadvertently gallop through them when they are rotated back into that paddock!

When to harrow:

- The best time to harrow a paddock is after it has been grazed and the animals have been moved on to the next paddock, unless this coincides with a period of heavy rain.

- Heavy rain will wash harrowed manure off a paddock more easily than undisturbed manure. So if heavy rain is forecast, wait until it has been and gone before harrowing.

- It is also preferable to mow any long grass that has been left by the horses in the paddock *before* it is harrowed, because the harrows will then distribute this around the paddock at the same time as the manure.

- After mowing and harrowing, the paddock should be rested until the grass has reached an average height of 15cm to 20cm (6ins to 8ins), at which point the horses can be allowed to graze it again.

Conserving pasture

Excess pasture can be conserved as hay, haylage or silage to be used later. There are many advantages to conserving pasture:

- It saves on the costs of buying in extra feed, including time and money spent on delivery/collection.

- It reduces the risk of weeds invading your pasture through bought in feed.

- If hay is kept dry (in a barn/shed) it can be stored for several seasons and although it will lose some feed value over time, it can be dampened before feeding as long as it is still 'clean'.

- Well packaged haylage/silage can be stored indefinitely without losing quality and does not require a barn/shed. This can help to shore off the effects of a future drought in countries prone to such climate conditions.

However, there are things to consider when planning on making hay, haylage or silage:

- It is not cost effective to buy the necessary machinery for a small property; most horse properties are not large enough to warrant owning pasture conservation machinery. It takes time to make and the harvesting season may clash with family or work commitments. Contractors are in high demand when you need them most; everyone's hay or silage tends to be ready to be cut and baled at the same time. What you can make will be dependent on the availability of contractors in your area.

- There is the chance that the weather can turn and the crop therefore spoiled, thus wasting any money spent on contractors etc. This risk is reduced with haylage/silage however.

- Taking forage off the land removes large amounts of nutrients, which must be put back either in the form of bought-in fertiliser or composted manure gained from the animals that ate the forage.

—

While on the subject of making hay for sedentary horses, it is also a good idea to cut hay on a cloudy day, because this will also reduce the sugar content of the final product. In fact, if it rains on the cut grass, as long as it is allowed to dry thoroughly before baling, this hay will be even lower in sugar as the rain will leach some sugar out.

The good news is that when hay manufacturers learn that what horse owners require is often the opposite of what general farmers require, they have a longer 'window of opportunity' for making hay. They can make hay for farmers when they are not making hay for horse owners and vice versa. This also means that if a horse owner relies on contractors to cut, turn and bale their hay, those contractors will be less busy when it is cloudy than when it is sunny, thus extending their availability. A win-win situation.

—

Excess pasture can be conserved as hay, haylage or silage to be used later.

'Foggage' ('standing hay')

There is growing interest in the practice of creating 'foggage' or 'standing hay'. This is where the pasture is allowed to grow tall and go to seed. Then, instead of being cut and baled, it is left (usually) until winter and grazed. Depending on growth conditions, it may actually still be possible to get one crop of hay, haylage or silage before locking the paddock up for foggage.

When it is time to use the foggage, it can be strip grazed for better usage; although the electric fence will need to be quite high if the plants are very tall.

Foggage negates all of the disadvantages of cutting and baling hay, haylage or silage, and there are various advantages:

- The plants have usually shed many of their seeds before they are grazed.

- The horses trample these seeds into the ground as they are grazing the foggage.

- There is no need to fill hay nets, cart hay etc.

- The longer plants protect wet soils better in winter.

- For horses that tend to get overweight easily, this may be the only time of year when they can graze for long periods.

- The yields for foggage may be lower than those for making hay, haylage or silage, but the elimination of costs completely negates this factor.

Things to keep in mind are that:

- Long grass is a fire hazard, however if it is well away from the house with fire breaks in between, or the surplus occurs at a low fire risk time of year, it may still be safe enough.

- This area will need to be mowed if it is not grazed evenly before the next growing season so that the new grasses can grow.

Bringing it all together

The use of various grazing management strategies offers great results for your land. If this all seems a little complicated, try to remember that rotational grazing is the main management strategy you should use. Pasture must have periods where it is allowed to rest and recuperate, therefore all land managers need to use this strategy so that paddocks get periods without any grazing pressure at all. Remember - the aim is to have healthy, vigorous plants, not stressed, worn out plants. The other grazing management strategies (e.g. limited grazing, strip grazing and cross grazing) are used in conjunction with rotational grazing to 'fine tune' grazing pressure.

There is growing interest in the practice of creating 'foggage' or 'standing hay'.

This is why it is particularly important that horse owners understand horses and pasture – as well as the management of that pasture. To simplify things, as stated at the beginning of this book, we need to mimic the natural behaviour of horses and grasses as closely as we can. This means utilising management systems for the grazing of our horses e.g., allowing periods of rest and recuperation in paddocks, encouraging or sowing lower NSC/less improved grasses and generally making sure that pasture plants are not 'stressed'. Knowledge is the key to taking good care of your horses so learn as much as you can.

—

See Appendix: ***The Equicentral System*** for information about a total management system that gives you even more control over grazing pressure and has big benefits for all concerned.

—

Chapter 3: Horses and water

Clean water is essential for healthy land and healthy horses. Pasture needs abundant water if it is to survive and thrive; without water, pasture will not grow at all. In turn a horse needs access to clean drinking water 24 hours a day. A single horse drinks between 5-50 litres a day which amounts to up to 19,000 litres of water a year. Water intake can vary depending on the ambient heat/humidity and the amount of work and subsequent sweating a horse does. Feed also influences water consumption, drier feeds require more water for digestion therefore horses eating green grass will drink less than horses eating hay.

Water is *the* most important element. As land owners and land custodians we *must* take care of the water system. We are part of that system and what we do affects not only our land, but the whole ecosystem downstream and even out to sea.

How water is collected, stored and distributed on the property is one side of the equation. So, you need to plan how to manage the water that arrives on the property. The other side of the equation includes the effect that horses and people have on the water system as a whole.

A single horse drinks between 5-50 litres a day which amounts to up to 19,000 litres of water a year.

Your water is part of the natural system

Fresh water comes from rain, fog/mist or melting snow. This water runs over the land via streams and rivers until most of it eventually ends up in the ocean. Some of this water soaks into the land, helps plants such as pasture to grow and adds to the groundwater aquifer (water table). Some evaporates back into the atmosphere to fall again as rain.

In land classification terms, the area of land that *catches* rain and snow is called a water catchment. In reality, we all live in a water catchment and the way in which we look after *all* of the water that passes through our land therefore affects both the local catchment area and areas way beyond.

As water runs over or drains from the land towards the main water courses (streams and rivers), it should pass through one or more riparian zones.

The Riparian Zone

The area of vegetation that surrounds, or should surround, any waterway, whether man-made or natural, is called the riparian zone. This zone should be made up of varying heights of plants and varying species of vegetation, such as aquatic plants, grasses, bushes and trees. Many species of plants and animals live in a healthy riparian zone. These areas usually support a high level of biodiversity and, if they are protected from stock, they quickly become thriving ecosystems. In these areas, it is common to see deeper soil (from silt deposits), native vegetation, more abundant plants, cleaner water and shelter for beneficial wildlife such as pest insect eating frogs.

Riparian zones are important because they:

- Filter and collect sediments and nutrients such as manure and fertilisers that are washed off the land, by trapping them in their foliage *before* they enter the waterway. They then use these sediments and nutrients for their own growth; a win-win situation.

- Buffer the negative effects of floods and wind by slowing them with their foliage.

- Act as habitat for wildlife, providing shade in the heat and shelter in the cold.

- Provide shade and a windbreak for humans, domestic animals and land.

A healthy riparian zone means that cleaner water enters the waterways.

The area of vegetation that surrounds, or should surround, any waterway, whether man-made or natural, is called the riparian zone.

The next stage

After passing through a riparian zone, the water then enters a stream or river and eventually travels out to the sea, or is captured in farm dams, lakes or wetlands first, where it is then held for a while before eventually moving on to the next stage (stream or river).

Wetlands in particular are vital resources because they act like a giant sponge, soaking up water in times of high rainfall and releasing it slowly into the next stage of the waterways; a stream or river. This reduces the incidence of flooding and erosion and is why areas that previously contained large wetlands, but have since been drained for development, often have flooding problems in times of high rainfall. Like riparian zones, wetlands also act as a natural filtering system by removing pollutants as they pass through the vegetation; the vegetation 'holds' particles and also uses them for growth when they settle. This means that when the water finally reaches a stream or river, it is much cleaner for having passed through a wetland first. Wetlands also provide vital habitat for wildlife. Ducks and other water fowl need places to care for young and shelter from predators, frogs need places to spawn, turtles need habitat and other animals rely on the inhabitants of wetlands for food e.g. birds of prey.

Wetlands in particular are vital resources because they act like a giant sponge, soaking up water in times of high rainfall and releasing it slowly into the next stage of the waterways.

Horses can damage this natural system

Poorly managed horses and land have a detrimental effect firstly on the local water system and then the water system at large (e.g. the water table, rivers and sea). The way in which you manage your land will directly affect your water quality and that of your neighbours near and far. Even if your land does not have or adjoin a waterway, what you do on your land will affect water further on in the system.

Poor horse property management results in bare, compacted and eroded soil. This causes problems far beyond the land in question:

- Areas of compacted soil repel rather than absorb water. When it rains the water runs over the ground rather than soaks into the ground and this runoff takes bare soil, manure and fertiliser into the waterways.

- This rainfall is not available to your pasture, because it runs off rather than soaks in and therefore reduces the pasture's ability to regenerate.

- Contaminants in the water cause pollution; even very small amounts of urine (ammonia) can be toxic to fish. Nitrogen and phosphorus are the biggest culprits for stimulating algae blooms and water weeds to grow.

- Excessive water weeds choke out the native vegetation causing death to these plants and to the aquatic life which survive on them.

- Dying plants and algae give off unpleasant odours and cause a drop in oxygen in the water which affects the fish's ability to breathe.

- Contaminated water is harmful for swimmers and for animals and people if used as drinking water.

- Loose soil and dried manure are blown away during dry periods. These are deposited in other areas, including the waterways.

- With some types of soil, rain can also wash pollutants into the soil to pollute ground water.

Poorly managed horses and land have a detrimental effect firstly on the local water system and then the water system at large.

If large grazing animals are allowed uncontrolled access to riparian zones and waterways, they damage them. They eat or trample the vegetation and cause compaction and erosion along the banks of the watercourse. Eventually, contaminants end up in the sea where they pollute marine life. All of these nutrients and organic matter should be going into your soil where they will be of great benefit - not the waterways where they cause devastation!

In addition, if horses are allowed to have direct access to water sources and riparian areas, they have a damaging effect on them:

- The trampling of horses causes soil (silt, clay and sand) to move around in the water, which can then clog fish gills, cover spawning beds, smother fish eggs and make it hard for fish to see their prey. This soil also coats in-stream vegetation and stops it from receiving the sun's rays. Mud is also created, which will then pass on further into the water system.

- Horses also compact the wet soils around the edge of the waterway which suffocates plant roots. The 'pugging' created by their hooves provides pockets of warm shallow water for mosquitoes to breed in.

- The hooves of horses fill in deep habitat holes that fish use to survive the heat of the day and hide from predators.

- Horses eat and trample seedlings and vegetation, which reduces the habitat for wildlife that relies on them. They also pull out plants, including the roots, which leads to bare soil and erosion.

- Erosion changes the course of streams and rivers, which leads to further erosion as the water moves faster rather than meandering, taking more soil with it.

Some of the signs that domestic stock has already caused damage to riparian zones and waterways are:

- Dying trees and saplings
- Short vegetation or a lack of vegetation
- 'Pugging' on the edges of waterways
- Scum and algae on farm dams and still water
- Smelly water
- Dying fish
- Lack of wildlife
- Plagues of mosquitoes

Therefore, allowing large grazing animal's access to water zones compounds the problem of poor management.

Creating and caring for a riparian zone

Clean water starts with either creating new riparian zones or protecting existing ones. This is done by fencing waterways and a surrounding riparian zone from domestic stock. Speak with your local land/soil conservation group, water catchment group, agricultural department or local authority about the recommended distance that the fence should be from the waterway, but 8m (26ft) is the minimum that you should be aiming for. You may be able to get either financial or physical help from certain groups.

Fencing off these areas enables the riparian zone to grow undisturbed. Initially, this fence can be a simple electric fence. Later, the area can be fenced with more permanent materials if you wish, but the sooner you fence the area the better. Take care to fence in such a way that native animals can still access the area.

It may be necessary to revegetate this area if it is damaged to the point that it cannot revegetate itself, or if weeds have completely taken over. If weeds are a big problem, care must be taken that their sudden removal does not result in further land degradation. You may have to remove weeds gradually by introducing more desirable species to the area. Find out and obtain the right planting list for your area by contacting a local native plant nursery or a land/soil conservation group. But remember you must fence first!

Trees should not be planted in farm dam walls, but can be planted further back where they will still provide shade for the farm dam. This reduces evaporation rates and the water temperature.

Clean water starts with either creating new riparian zones or protecting existing ones.

Before...

After...

Once fenced (if safe to do so), the area within the fence could be used for *very occasional* grazing. You need to wait until the area is well vegetated. So, make sure you plan to fence as large an area as possible and include a gateway. The fence will allow you to completely control the amount of grazing that the area receives; if this area is left unfenced, it will be grazed too short for a healthy riparian zone to become established.

If animals need to cross a stream or river, then plan to build a crossing. A crossing should be fenced on either side and concreted or covered in rocks; a permit may be required for this so check before building. Keep in mind that fast flowing water, which can occur infrequently but without much warning, can wash away fences and stones etc.

If animals need to cross a stream or river, then plan to build a crossing. A crossing should be fenced on either side.

If necessary, *limited* access to waterways can be created by fencing a u or v shape into the stream or river. This area could be covered in stones and gravel to reduce pugging and silting. Ultimately it is much better to prevent access altogether. This will not work if the river is fast-flowing either some of the time, or all of the time, as the fences and stones etc. will be washed away. If necessary, the water can be pumped (check with the local authority if it is a stream or river as you may need a permit) to troughs or holding tanks and distributed from there.

Once restored, this area will be a far more beautiful feature on your land and will operate far more efficiently than simply providing grazing for stock.

A fenced farm dam is also safer. Depending on the type of fence, it will keep small children out – therefore a farm dam that is situated anywhere near where small children are likely to gain access should always be fenced. A farm dam can also be dangerous to horses and other animals if it becomes silted up and the water evaporates (during a drought). In this case the farm dam becomes a thick soup of dangerous mud and animals can become stranded and can even drown if it is their only water supply.

When you have protected the riparian zone in this way, the result is much cleaner water which can then be reticulated (collected and distributed via tanks, pumps and troughs) to your stock when needed. A healthy riparian zone also reduces water loss through evaporation, meaning your clean water lasts longer.

Conserving water

Saving small amounts of water adds up, leading to a reduction in overall demand. As well as protecting waterways, there are many things that you can do to conserve water so that less water needs to be taken from the water system as a whole.

Some of the things that you can do to conserve water are:

- **Improve the water holding capacity of your land.** One of the aims of good land management is to slow down the rate at which water *leaves* the land. Swales can help to slow water which means more water is absorbed into the land (see the section *Using swales*).

- **Sponge horses rather than hose** to reduce water usage. Aim to wash horses less often and groom them more often; they prefer that anyway. If you do hose them, do it on a grassed area or have the water drain into a system for recycling such as a grey water system already attached to the house, or a separate one for the horse facilities area. Use only environmentally friendly products that are mild and biodegradable for washing your horses.

- **Either set up your property so that it does not rely on irrigation, or use irrigation carefully if it is already installed**. There is no point in irrigating bare, unimproved paddocks that are unable to soak up and use the water. Likewise, there is no point in watering grasses that are not currently in their growing season (see the section *Irrigation*).

- **Water garden plants with trickle systems or micro irrigation** and only water in the early morning or late evening. Adhere to your local authority's water restrictions.

- **Reduce lawned areas** if they need watering and convert them to pasture. Well managed pasture is just as beautiful and more functional. Mulch plants to reduce water evaporation (this also keeps them warmer in winter). Mulching around the base of plants suppresses weeds, acts as a slow release fertiliser for plants (depending on the mulch type), provides an environment for plant friendly organisms and reduces evaporation.

- **Ensure taps and auto drinkers do not drip** (including paddock drinkers).

- **Be careful not to allow troughs to overflow** when filling them.

- **Collect as much rainwater as possible**; All buildings should have water tanks attached (see the section **Water storage tanks**).

- **Speak to your local land/soil conservation group** and find out who can help you with water collection and conservation strategies.

Collect as much rainwater as possible; All buildings should have water tanks attached.

Water problems

Common water problems on a horse property include there being too much water or not enough. Water usually arrives all at once and then not at all for very long periods, especially in countries such as Australia and parts of the USA. By managing water correctly, these booms and slumps can be ironed out to some extent.

Water should move slowly across the land, giving the soil and plants time to soak it up and use it. Water can be slowed down with mulch, the use of swales, fences that follow contour lines, correct placement of vegetation and many other methods (see the relevant sections).

Also, keep in mind that areas which are surfaced create much more water run off than areas that are not (water seeps into areas that are not built on).

It is just as important to manage water that stays too long on your land. Waterlogged soil and its subsequent problems will result if you continue to graze pasture that is too wet. Somewhere to 'hold your horses' is just as important in areas that are too wet as in areas that are too dry (see Appendix: *The Equicentral System*). It may be necessary to get the advice of local drainage experts, although often a change in the horse management can have big results.

Impure water

Water can contain many impurities; bacteria, fungi or excesses of minerals, all of which can be tested for. If you have any doubts about the quality of water on your land, you must have it tested before using it; your local authority may be able to help you with this.

Fencing off a water course and allowing vegetation to grow around it helps enormously to keep the water cleaner, see the section *Creating and caring for a riparian zone.*

Water can contain many impurities; bacteria, fungi or excesses of minerals.

Sources of water

Water primarily arrives on your land via rain that falls on it directly, and as rainwater running neighbouring land, which travels across your land towards a body of water such as a farm dam, lake, wetland, stream or river.

Other potential water sources on a property include:

- Mains water, which is water that is piped to your property by the local authority in urban and some semi-urban areas – but not usually rural areas depending on which country you live in.

- Well (bore) water, which is water that is pumped up from the underground water table (aquifer).

- Natural spring water, which is water that naturally rises to the surface (from the underground water table (aquifer) in certain areas.

- Streams or rivers that run through or alongside the property.

This photo may look idyllic at first glance, but look at the bare soil and therefore lack of riperian area around this waterway.

Mains water

Different countries have different arrangements for 'mains water'. In the UK for example, most properties are connected to the mains with only 1% not being connected. Therefore, even many rural properties have mains water. By contrast, in Australia and New Zealand, rural properties rarely, if ever, have mains water. Only those that are near to suburbia tend to have it. In the USA and Canada, a similar situation occurs with 15% and 25% of homes respectively not having mains water.

If a horse property *does* have mains water, then it may be metered and an added expense to use for watering stock. Consider installing rainwater storage tanks on any buildings that can then be used for watering animals. These are becoming more popular even in countries that do not traditionally use them (see the section **Water storage tanks**). Ideally, a property would border a running waterway and have permission to pump that water.

Well (bore) water

Well water is water drawn up from a natural water aquifer underground, often called the 'water table'. In modern times, a 'bore' hole is drilled until underground water is found and this water is pumped up to the surface by an electric pump . Before electric pumps were used, water was drawn up to the surface physically; hence the old fashioned image of a well being a covered circular wall around a hole into the ground and a winding mechanism for the bucket.

In some areas, the water table is huge and contains abundant high-quality water. In other areas, it is either non-existent, limited or of poor quality. Worldwide demand for water is enormous and in many areas these underground stocks of water are dwindling through overuse. Be aware that well water might contain minerals that can be harmful to stock and fauna. It can be very high in salt in particular and can therefore cause a lot of damage to the land. If you have a well on your property always have its water analysed before use. Seek advice from the relevant local authorities before constructing a new one.

Aquifers vary in how close they are to the surface. Well water is exhaustible and in some areas it is being used faster than nature can replace it so causing new environmental issues. If your property does not have a well but you are thinking of putting one in, you need to weigh up the costs with the benefits. In many areas, there are now restrictions on putting down both domestic and irrigation wells.

Natural spring water

To have spring water is obviously highly desirable as it should mean that the property will always have water. Some spring water can be quite heavily mineralised, which may make it unpalatable at least initially until the animals become accustomed to the taste. Like well water, it should be tested before using for stock. A farm dam or lake may be able to be created around a natural spring which results in a constantly full waterway.

Streams and rivers

Domestic animals should not have direct access to streams and rivers for use as drinking water due to the extensive damage that they can cause to the ecology of the waterway. Also, if a waterway has a sandy base, the horses will take up this sand each time they drink, potentially leading to sand colic over time due to an overload of sand in the gut. In addition, water that is not moving becomes stagnant and this water is then not clean enough to drink. If the property has a licence to harvest water from the waterway, it is better to pump the water to one or more holding tanks that are placed higher up on the property (called a header tank – see the section **Water storage tanks**). Gravity will then allow this water to be reticulated to water troughs below this level.

*Domestic animals **should not** have direct access to streams and rivers for use as drinking water due to the extensive damage that they can cause to the ecology of the waterway.*

Storage of water

All properties have the opportunity to collect rain water both directly and as run off. If you live in an area that tends to run out of water at certain times of each year, then capturing this water and storing it in tanks and farm dams for later usage is an efficient way of ensuring water self-sufficiency.

Water storage tanks

Rain water from the roofs of buildings can be collected and stored by fitting gutters and spouts to all buildings that run to water storage tanks. Water storage tanks can be concrete, corrugated steel or polyurethane. These tanks are very common on rural properties in Australia and New Zealand, but are only just becoming available in other countries. If you live outside Australia or New Zealand and you are interested in harvesting and storing water sustainably, do some research and see if they would prove to be cost effective in your locality. To get started, do an Internet search for **Water storage tanks**.

Rain water from the roofs of buildings can be collected and stored by fitting gutters and spouts to all buildings that run to water storage tanks.

Any overflow/excess water should be channelled to a farm dam/lake or other waterway without having to go through a polluted area first (e.g. through horse holding areas or past manure storage areas). Plan to create an area that has a high level of vegetation for this runoff water to pass through before it enters a waterway, this vegetation will help to filter out any nutrients and pollutants. If you have excess water on a regular basis, then there may be a case for putting in more water storage tanks.

Water can then be reticulated from the storage tanks to where it is needed on the property (see the section *Reticulating water*).

If the water in the tanks runs out and you rely on this as your main water supply, then water will need to be bought. This will need to be delivered by water tanker and it works out relatively cheaper to have a larger tanker deliver the water as most of the cost is associated with the transport. Make sure you plan to have space and adequate surfacing for water trucks to enter and turn around on the property.

Farm dams (ponds/lakes)

Ranging from man-made to natural, these areas can be a real boon on a horse property if managed well. Farm dams are another valuable way of catching and storing rain water that would otherwise run off the land. A farm dam can be sunk into the ground on all sides or be formed by retaining walls of compacted soil (on the side that water exits the farm dam), either way farm dams should preferably be shaded to reduce evaporation. However, trees that will develop large roots should not be planted into or near the dams retaining walls. Root growth or death can threaten the integrity of the dam's wall and lead to water leakage over time.

As well as being a useful supply of water for stock, a farm dam can be an attractive recreation area and provide habitat for wildlife. Some of this wildlife (such as insectivorous birds and insectivorous bats) will be beneficial to the property because they eat pest insects in particular. A farm dam can also be stocked with fish, some of which will eat the larvae of mosquitoes; seek qualified advice about what type of fish to stock it with – make sure you are not introducing a pest species. Frogs eat lots of pests too, so encourage them to live in your farm dam by creating inviting areas for them with rocks and suitable fauna.

On a very small horse property, you need to weigh up whether a farm dam and its riparian zone will take up too much space that could otherwise be used for pasture. In this case, it may be better to spend the money on extra water collection tanks.

If you plan to have one or more farm dams, keep in mind that if space is limited, it is generally better to have one large farm dam than several small ones, because small farm dams are not as effective. In a hotter climate, a farm dam needs to be large and deep (at least 4m/13ft) in order to keep the temperature low and reduce evaporation. Small, shallow farm dams stagnate sooner due to increased evaporation rates.

A well planned and managed farm dam is fenced off from stock, so it is not necessary to have one in each paddock. Therefore, one large, strategically placed farm dam will work much better. For example, a farm dam can be placed in a high position on the land from where it can collect rain **and** water from land that is higher still. There is potential here to spare resources by using gravity to feed the water to where it is needed. Again, get qualified advice about the positioning of a farm dam before construction. In areas that have problems with salinity (some parts of Australia and the USA for example), a large body of water in the wrong position can create or increase an existing salinity problem.

These areas can be a real boon on a horse property if managed well.

If the farm dam is to be used for reticulation, then water should be taken from the middle levels. If water is drawn from the top 20cm (8ins) of a farm dam, it is more likely to be contaminated with potentially harmful micro-organisms. However, water taken from near the bottom of a farm dam will be colder and lower in oxygen due to the action of micro-organisms that use oxygen to break down any organic matter at the bottom of it.

If a layer of algae builds up on the surface of a farm dam, you can put a few bales of hay (still in their strings) into it. These bales will float and the algae will stick to them. They can be removed after a few weeks and used as mulch or put onto the manure pile. As well as helping to clear the algae, they also provide a floating platform for wild birds.

Periodically, just before or during a wet period, it is also a good idea to siphon or pump water from the bottom of the farm dam out on to the paddock. This results in fresh, oxygenated water replacing the water from the bottom of the farm dam, which is low in oxygen.

Check with your local authority before having a farm dam constructed as you may need permission. As already mentioned, expert farm dam construction advice should also be sought before construction; some soils, such as sandy soils with no clay content, will not hold water. Clay can be brought in, but it adds greatly to the expense. The farm dam must be compacted properly with heavy machinery when constructed and the spillway should be placed properly in order to be effective. If a farm dam fails and causes damage to a neighbour's property, you may be liable. Therefore, a farm dam should be installed by a qualified and insured earthmover to reduce the chance of failing and it must be managed properly if it is not to become a stagnant, toxic pool.

Using water

At the planning stage, you need to consider how you are going to get water to the various parts of the property (reticulation) and whether you are going to water parts of the property (irrigation).

If you live next to a waterway and intend to use this for watering your stock, it may be necessary to pump the water up to holding tanks and distribute it to water troughs from there. This would protect the stream bed and bank from trampling and provide clean, healthy water for your horse/s.

If you intend to use water from the river or stream to irrigate pastures, then you may need to apply for an irrigation licence (your local Council should be able to assist). If it is necessary to install a water crossing, this should be fenced on both sides and the base concreted or in-laid with rocks to avoid further land degradation. You may need to seek permission from your local authority in order to undertake any work on a watercourse.

Reticulating water

Water can be reticulated around the property from the water source (mains, well, holding tanks etc.). This can be done using polyurethane pipe ('polypipe') that runs to troughs, automatic drinkers in stables and surfaced holding yards, and also to sprinkler systems if required. If the water is from holding tanks, you will need an electric or solar powered pump.

If the land is undulating, an extra tank can be positioned in a high area and water from the tanks near the buildings can be pumped to this (header) tank. The higher up a hill this tank is positioned, the more water pressure there will be. This tank can then gravity feed troughs etc. Using a header tank also creates more space in the tanks near the buildings so that next time it rains they can fill up again. This system also means that instead of the pump having to start up every time you turn on a tap, it only has to operate occasionally to pump the water up the hill to the header tank. Another possible option involves pumping water from a farm dam/lake to a header tank and then using this water for the toilets, washing horses, watering stock etc., saving the cleaner tank water for drinking and washing in the house (if the house does not have 'mains' water).

If you are relying on an electric pump, it is also a good idea to have a petrol pump that can quickly be attached in the case of a fire. This is most important in a country that is likely to have a bush fire; when a fire occurs, the power supply is often the first thing to go down, rendering you unable to fight the fire. A portable petrol pump is useful in that it can also be moved around the property if necessary.

If using polypipe, think where it is going to run, usually underground or along fence lines. If you need to excavate trenches, it is better to only hire the machinery once and it may also be easier to do before any permanent fences are put in. Plan where you might need taps etc. and have connection points put in at this stage. You may need to isolate certain areas for repairs etc., so plan where to put stop valves.

The diameter of the water pipe leading into the trough should be of sufficient size to allow refilling of the trough in an acceptable time. Typically 32mm is a starting point in a non-mains pressure type reticulation system.

Water troughs

Avoid putting a water trough near a gate, as they create even more wear and tear in this area, or in valleys, as horses will traffic downhill to them, creating the conditions for erosion. Low lying areas tend to be wet already, so a trough placed in a dip will tend to create a muddy area.

Also, avoid placing a trough in the corner of a paddock where a horse may be in danger of being trapped by another horse.

Troughs can be placed so that they can be accessed from two paddocks to save costs, unless you have a double fence and then the trough will not reach both areas.

A trough should be set on a level sand base and surrounded by a firm surface that will resist hoof action. Place troughs in an area where the ground drains well in wet weather and place gravel around the trough to reduce mud and dust.

Because this area will be bare of grass, also consider placing a trough on the windward side of the stables/surfaced holding yards or other facilities, where it will protect the area against bush fire if there is risk of this occurring.

A reticulated water supply usually relies on a float-valve to control the water-flow into the trough. If the float valve is a type that horses can tamper with, it should be covered to prevent this happening.

A trough can be made from concrete, plastic or steel and their sizes can range from a small plastic or metal 'automatic drinker' usually situated in a stable or surfaced holding yard, to a larger trough for a paddock. A round trough with a one metre diameter or a bath sized rectangular trough is ample for a small group of horses. Keep in mind that some horses will try to splash around in a large trough.

An old bath can be used as a trough on a horse property, particularly if the water supply is not automatic (e.g. just using a hose attached to a tap) or again a float can be fitted if you want reticulated water. Baths are easy to clean and are fine for horses as long as they have no sharp edges that will cause injuries.

Whatever you use to supply the water, it and its access must be checked daily (even more often in hot weather), as it can be knocked over, become contaminated (e.g. by a dead rodent/bird or manure), the horse/s may not be able to get to the water for some reason (e.g. someone accidentally shuts a gate which prevents entry) troughs can break down, streams can stop flowing and water can freeze in cold areas. Horses cannot break ice on troughs and will die of thirst in this situation.

—

This is another advantage of **The Equicentral System** – there is only one water trough to check and it is best placed near the house – see Appendix: *The Equicentral System*.

—

Avoid putting a water trough near a gate.

Irrigation

Irrigation is not essential on a horse property and it is quite possible to manage without it. Installing irrigation can be expensive both in monetary and environmental terms. Changing the pasture species to ones that manage better with less water and improving water retention on the land by improving the water holding capacity of the soil/reducing compaction are more viable options. In fact, you should always be looking to improve the soil.

Water for irrigation comes from farm dams/lakes or ground water via a well; mains water is not usually an option due to the expense.

Irrigation is installed for many reasons including:

- The growing season for pasture can be extended. For example, if rainfall is a limiting factor but there is enough sunshine, irrigation will allow plants to still grow.

- Plants that would not usually be able to be grown can be. For example, in a climate that has hot dry summers, plants that are adapted to wet hot summers can also be grown.

- Pastures look better; greener pastures are more aesthetically pleasing. This is after all why people keep 'lawns' green.

- For growing commercial crops e.g. crops or Lucerne hay.

If the property already has an irrigation system in place, your main concerns are whether the system is operating efficiently without water wastage.

There may still be things that you can do to improve the system:

- Make sure the system is not putting more water on the ground than it can cope with. A maximum of 10mm (0.4ins) per application is recommended to avoid fungal problems and waterlogging, however if the soil is compacted any amount of water will run off without soaking in.

- Ensure the plants are of a type that will benefit from irrigation. You need to either sow pasture plants that will benefit from the water, or do not water. There is no point in watering plants that will not grow at that time of year or in watering a paddock full of weeds. Incorrect irrigation can actually bring on more weeds.

- Keep in mind that some plants cannot cope with the higher levels of salt sometimes present in well water.

- Carry out good pasture management (e.g. paddock rotation, manure management etc.) as for non-irrigated pasture. Rotations will need to be more frequent due to the quicker herbage growth.

- Aim to be careful with the water, whether it is from a dam/lake or a well and minimise usage by watering at the right times; do not irrigate during the day in hot weather - early morning or later in the evening is best. Use timers so that drinkers do not get left on.

- Check and maintain the system before a dry period/time of year. Protecting sprinklers from horses and freezing temperatures can be difficult but not impossible.

- Prevent water from irrigators spilling onto roadways or other un-pastured areas. This is wasteful and understandably upsets neighbours.

If the property does not have an irrigation system in place, you need to weigh up the cost of installation with the costs of sowing different pasture and the long term benefits to you of both options. It may be that it is not worth the extra expense taking into account the chances of failure of the system e.g. if the well stops producing water, compared to the costs of using surfaced holding yards etc. for the horses and buying in extra feed.

Pasture can usually be vastly improved without irrigation. Even when the shortfall in feed has to be made up by purchasing feed, it may still be far lower than the cost of irrigation (including the environmental and labour cost). You should also employ strategies to ensure that any water arriving on the property (e.g. rainfall) remains on the property for as long as possible, especially if you live in an area that has long dry periods. Improving the water holding capacity of your land is something that should always be done before irrigation is installed and this may be enough in many cases.

If you plan to install a new system, aim for ease of use and economy. In small paddocks, the system can be positioned around the fence line. A portable system using hoses and free standing sprinklers can be used for the middle sections if necessary. In larger paddocks, it may be necessary to position permanent sprinklers in the middle of the paddock, as well as around the edges. Popup sprinklers work well with horses as they can sink underground out of harm's way when the paddock is in use. Post fastened sprinklers are protected from horses if they are currently grazing the paddock. If placed high above the ground (e.g. 1.8m or 6ft), they seem to get less damage from horses than if lower. Where possible, these should not cause obstructions to machinery for paddock maintenance.

If you are thinking about installing irrigation, it is a good idea to talk to people who already have it to get some ideas. Some irrigation suppliers give a design service, however this will be aimed at using all of their products whereas it may be more cost effective to use products from various outlets. Make sure the design is not more elaborate than you need, but if you plan to expand now is a time to consider your needs of the future.

Planning for clean water

It is a good idea to work out a water management plan before you fence a property.

Things to keep in mind are that:

- Do not use steep hillsides to paddock horses, as the soil and manure is washed more readily into the water system.

- Horses on steep land also cause erosion which leads to more soil ending up in the waterway.

- Water run-off from slopes should be directed *across* the landscape rather than directly down any hills when possible. This slows the water down, reduces the damage that it can do and gives it more opportunity to soak into the ground.

- This process can be helped by fencing paddocks and laneways along contour lines whenever possible and keeping any fences that go downhill as short as possible.

- Any streams and rivers that go through or border the property (and any farm dams), should be fenced off and if necessary the water can be reticulated around the property.

- Don't allow bare areas to form anywhere on the property, bare soil will be blown or washed away and is usually deposited in the waterway. They are also an invitation for weeds.

- Take care with fertilisers, either chemical *or* organic. Even though a large amount may be recommended, it may be better to apply it in stages, so that any run off will be reduced. This is a 'catch 22 situation', as fertiliser, correctly applied, will help plants to grow vigorously, which in turn will vastly reduce the flow of pollutants into the waterway by trapping and using the nutrients. Correct fertilisation is important, as is the timely application of that fertiliser. Remember, it is usually possible to improve your soil with minimal use of fertiliser; ensure that you are only using what is needed.

- Install rain gutters and either collect water in tanks or divert the run off so that it does not run across holding yards and manure storage areas.

- Site any new buildings and holding yards as far from a waterway as possible; aim for at least 100m, again you will need to check with your local authority, and on the highest parts of the property. Make sure that contaminants do not leach into the soil by siting structures on an impervious layer such as concrete or compressed limestone.

- Use environmentally friendly biodegradable products for washing horses and washing horse gear; some shampoos actually prevent water from entering the soil.

Use environmentally friendly biodegradable products for washing horses This horsewash contains footings of gravel which help to filter out any impurities after washing and before the water is used to irrigate the pasture.

Chapter 4: Horses and vegetation

A horse property should be a biodiverse haven containing many different species of vegetation. Creating this is one simple way to transform people's perceptions about horsekeeping from negative to positive and help them to understand that a horse property *can* be an environmental asset.

This section is concerned with vegetation *other* than pasture. Aim to grow herbs, succulents, bushes and trees in as many areas as possible for the huge benefits that they provide.

The main points when considering what types of vegetation to plant, protect and/or encourage on a horse property are:

- Which species are likely to be safe in terms of poisoning?

- Which species are not going to cause safety issues in the future in terms of dropping large limbs, being a fire hazard or being difficult to maintain?

- Which species are not going to attract problem species of insects or animals?

- Which areas on the property would be best to put aside for vegetation other than pasture?

- Which species are considered local weeds and why?

- Which species will enhance the ecosystem of the property?

A horse property should be a biodiverse haven containing many different species of vegetation.

Some of the benefits of trees and bushes

Trees and bushes can carry out various functions. The following is a list of many of their benefits on a horse property, most of which are expanded on later in this chapter. **With careful selection and placement, trees and bushes can do all of the following:**

Help with temperature control:

- Trees and bushes cool a property in hot weather by providing shade for people, animals and buildings. They slow down hot, drying winds, reducing the reflective heat, and cool the air temperature because of *transpiration* -the process of water movement through a plant and evaporation through its leaves.

- They warm a property in inclement weather by providing shelter for people, animals and buildings and they slow down cold winds and driving rain.

- Therefore, trees and bushes balance out some of the harsh extremes of a climate. This benefits everyone because of lowered stress which can be caused by excessive heat and reduced energy wastage (caused by trying to keep warm).

Trees and bushes cool a property in hot weather by providing shade for people, animals and buildings.

Help with safety and biosecurity:

- Trees and bushes planted between paddocks reduce fence walking behaviour of horses in adjacent paddocks and prevent horses from playing over fences, therefore reducing fence injuries.

- Trees and bushes on the boundary of a property slow down the spread of diseases from one horse property to another by providing a physical barrier to pathogens in the air and by preventing horses from neighbouring properties from touching each other.

Provide cheap fencing:

- Trees and bushes are also called 'living fences' by some people. They can improve an existing fence or can be incorporated into a new fence. Either way, the result is a cheaper, better fence that has many benefits other than simply dividing a paddock.

Trees and bushes on the boundary of a property slow down the spread of diseases from one horse property to another.

Help with fertilisation:

- Trees and bushes bring up nutrients from far below with their deep root systems and store them in their leaves. These nutrients are then deposited on and in the top soil when the leaves fall.

- They attract animals, for example everything from birds to large mammals, resulting in more manure being dropped in the area.

- Trees situated high up on a property have the benefit of spreading nutrients from the leaves and the manure from the animals to fertilise the land below.

- Some species (legumes) 'fix' nitrogen from the air into the soil, thus helping other plants and the grass to grow whilst reducing the need to add fertiliser.

Help with fire control:

- They can slow down fire; by planting the right vegetation in the right place fire can be slowed down.

Help with drainage and erosion control:

- Trees and bushes can slow down fast moving water if they are planted in the correct place. Their roots also absorb a lot of water, therefore some particular species of trees and bushes can be used to dry out a waterlogged area. The roots of trees and bushes also hold soil together and therefore reduce erosion. See the section *Easy areas to increase vegetation*.

Help with weed control:

- Trees and bushes can slow down the spread of airborne weed seeds, again by providing a physical barrier. They can also inhibit certain sun-loving weeds by shading them out.

Help with pest control:

- Trees and bushes provide habitat for wild animals, which helps to create a natural, diverse ecosystem rather than a mono culture. Variety of wildlife provides huge benefits, for example, certain birds and bats are primarily insect eaters and help to control pests insects such as mosquitoes, flies, caterpillars, grasshoppers and aphids. They give natural pest protection; certain plants will repel insects. If you are interested in this subject, more information can usually be found in any local library.

Add value to a property:

- Trees and bushes add value to a property by creating an attractive environment in which to live, by giving privacy from neighbours and by reducing noise pollution.

Provide feed for stock:

- Trees and bushes provide feed for stock. In some parts of the world, trees and bushes (rather than pasture) sustain animals. If you are interested in this subject, you will need to research which plants will not become a problem for your location in the future. What constitutes a good fodder tree in one country (or even one part of a country) can be classed as a noxious weed in another.

- Some of these plant species will have health benefits for horses, some of these benefits are already known and others are yet to be discovered.

Reduce the sugar levels of the pasture:

- Large vegetation shades a paddock so that less sunshine gets to the pasture plants. This results in the plants being lower in sugar and starch. In the case of horses that are at risk of laminitis or other metabolic and/or obesity related disorders this is a plus point that can help in their management.

Offset carbon:

- By nurturing flora, you are assisting in the reduction of carbon based pollutants in the atmosphere. Vegetation takes carbon from the air and deposits it in the soil.

Help to keep the waterways clean:

- Vegetation growing next to waterway shades the water and reduces evaporation from the waterway when the weather is hot. This cooler water can then be reticulated around the property as and when needed.

- Vegetation also 'filters' nutrients, sediment etc. that are in the water. The plants use the nutrients to grow and the water that passes on to the next stage in the system is cleaner.

Reduce stress/increase mental health:

- Green spaces have a calming effect on humans.

Provide an additional resource/income:

- Some trees can be harvested or provide firewood; some major companies will even pay landowners to plant trees to offset their carbon emissions (called 'carbon offsetting').

Vegetation growing next to waterway shades the water and reduces evaporation from the waterway when the weather is hot.

Trees and bushes as habitat for wildlife

Horse property owners can benefit the environment hugely by providing habitat for wildlife. Trees and bushes are habitat for numerous species of insects, mammals, marsupials (in Australia) and birds. Any existing wildlife on the property needs to be protected and habitat created for other different species. The more varied the ecosystem, the more sustainable the property will be. This does not have to be at the expense of grazing land, as maintaining and creating habitat for wildlife has numerous benefits to the property and the domestic animals (and people) that live on it as a whole, **some examples are:**

- The pest control benefits mentioned earlier.

- Species of native beetles, spiders, centipedes, bees and other insects in turn all have an ecological niche and play an important part in maintaining a healthy ecosystem.

- Certain small mammals and marsupials eat insects, including those that damage plants.

- Small insectivorous bats also eat many insects, such as mosquitoes, by the thousand on a daily basis.

Encouraging and protecting these beneficial creatures involves creating and protecting their habitat. This can be done by fencing off areas of existing vegetation, or creating new areas (such as areas described above). Make sure that any fencing allows native animals to pass through – preferably linking these areas to others both on the property and neighbouring properties to create 'wildlife corridors'.

Find out which birds and other beneficial animals or insects migrate to or live permanently in your area and encourage them by planting trees and bushes that attract them. Old, hollow trees and logs should be left alone, as native animals rely on them for habitat. In fact, you may need to look at providing man-made habitat if there are no natural examples around, as it takes many years for these to develop naturally. For example, certain birds rely on old trees that have the hollows that they need to nest. Rocks provide habitat for frogs. Speak to your local land/soil conservation group for advice.

Horse property owners can benefit the environment hugely by providing habitat for wildlife.

Fodder trees and bushes

In some parts of the world, trees and plants (other than grass) sustain animals. Grasses generally have shallow roots and cannot survive in very harsh conditions. There is a growing interest in feeding domestic animals with fodder trees in western countries due to the success of fodder trees in countries with harsh conditions, especially desert terrain.

Most fodder trees share certain characteristics such as:

- They have long roots so that they can access water from deep below the surface and therefore provide fodder even in very dry conditions.
- Many fodder trees provide more feed than pasture per square metre of land.
- Many are legumes; therefore they fix nitrogen in the soil reducing the need for nitrogenous fertilizer.
- Many have high protein yields.
- Many fodder trees are also good windbreak and fire resistant trees.
- They can have health benefits, many of which are currently being researched.

When sourcing fodder trees, opt for local, native trees whenever possible. You also need to take into account your climate and soil conditions.

Fodder trees and bushes need to be pruned or grazed regularly so that they do not grow too tall. Aim for no more than 2m in height so that both you and your animals can reach the top. Do not plant them as boundary plants, as they need to be able to be accessed from both sides, therefore internal laneways/driveways and corners of paddocks are best. They can also be planted around an all-weather surface or training yard and on the outside of surfaced holding yards. You may want to plant them as an orchard of trees and allow the horses in for periods of grazing. Wherever you plant them, a variety of species is a good idea so that horses can browse as they would naturally. The plants need to be alternately grazed and rested in just the same way that pasture does. They will need to be fenced off from stock for at least the first year after planting (maybe longer) so that they can become established.

Trees and bushes provide feed for stock. Bushes such as the Tagasaste (tree lucerne) make an excellent supplementary feed source, but please check with local authorities before planting to ensure it is not declared a weed in your area.

Windbreaks and firebreaks

Trees and bushes as windbreaks help to stop moisture loss in soil (by approximately 20% – 30%). They reduce evaporation from the soil, enabling the plants to utilize the moisture more effectively and increasing growth periods and therefore yield. They also reduce erosion by wind on dry soils.

They need to be across the line of the prevailing winds to be effective; plan to plant shelterbelts that will benefit areas such as surfaced holding yards and paddocks as well as a house and other buildings. Double fences and laneways between paddocks provide an excellent area to plant windbreaks.

A good windbreak slows the wind down, protecting everything on the lee side. Windbreaks should be constructed so that they *slow* the wind speed by 60% to be effective. Solid windbreaks cause the wind to gust over them and create turbulence on the lee side. Permeable windbreaks are better because they simply slow the wind as it goes through. A good windbreak will extend its effect 20 times its height downwind and 5 times its height upwind of the windbreak. For a shelterbelt of trees to do its job correctly, it should be at least 20-25 times as long as it is high. The trees need to be stepped by planting in at least three rows with the tallest trees in the middle. This is achieved by planting different species that have different heights at maturity.

If fire is a risk in your area/country, windbreaks should also be designed with this in mind. You need to know what direction fire is likely to come from and plant at right angles to this direction.

They should not be positioned too close to buildings or surfaced holding yards. In particularly high risk areas where the windbreak is too close to buildings, remove the middle story of plants to reduce the risk of bushes igniting the tree canopy.

Even fire resistant trees and bushes will burn eventually, however they do offer some protection. They reduce the wind speed and this reduces the speed and intensity of the fire. Trees absorb the radiant heat, protecting animals and buildings on the lee side. They also catch airborne burning material, stopping it from flying into other areas and starting spot fires. Organic matter around the base of these trees and bushes needs to be kept clear as this counts as fuel.

Aim to plant flame resistant varieties of trees if fire is a risk in your locality; flame resistant trees and bushes tend to have certain characteristics in common. These characteristics are:

- They are usually slow growing (e.g. deciduous trees).
- They usually have a high salt content (such as mangroves and salt bushes).

- They should not collect litter within the canopy or on the bark (smooth barks rather than stringy or paper barks).

- They usually contain high moisture levels in their leaves (rainforest plants, cacti, succulents and most fruit trees).

- They usually contain low amounts of volatile oils.

Always seek advice from your local fire authority regarding any use of trees/bushes with regard to fire prevention.

Every area differs in what can be grown and in what is available. A local land/soil conservation group will usually be happy to provide you with the names of useful shelterbelt and firebreak trees for your district.

A good windbreak slows the wind down, protecting everything on the lee side.

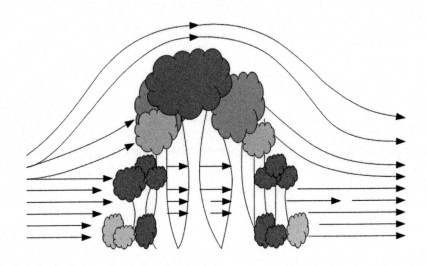

Revegetation of steeper land

In many cases, land is sold for horsekeeping that is actually too steep for grazing of large animals. Steep land is far harder to manage well than is the case with flatter land.

Steep land can be problematic for the following reasons:

- Weed management is more difficult because it is not safe to drive machinery (such as a mower) on steep slopes. Manure is also more difficult to manage for the same reason, because it is not safe to drag a pasture harrow on steep slopes.

- Manure is more readily washed into the waterways on steep land.
- Gravity and water (rain) result in soil being washed downhill. This process is exacerbated if the land is overgrazed and therefore there is not enough vegetation to hold the soil together.
- Large grazing animals add to this problem (in addition to overgrazing if this is allowed) by physically 'pushing' the soil downhill. This speeds up the process of soil erosion and soil being washed into the waterways.

In addition, horses should not be permanently kept on steep land as this will lead to joint problems over time. Horses are fine grazing up and down hills of course, but they need to be able to get to flat areas to 'loaf' (sleep and 'hang around' together). This is because they are not comfortable/cannot sleep either standing in that characteristic position with one leg resting or lying flat out, on a hill side. Horses have virtually no sideways movement in the joints of the legs, so forcing them to stand on the side of hill puts an unnatural force on these joints which over time will lead to joint problems.

Sometimes it is easier, and much better for the environment *and* your animals, to fence off any steep areas on the property, re-vegetate these areas with trees and bushes, and concentrate on growing pasture on the flatter areas of the property. This will help to fertilise the flatter ground lower down on the property.

Trees and bushes on hillsides and the tops of hills help with fertilisation of land below them by attracting a huge variety of insects and animals (and their dung). The leaves also aid fertilisation and the plant roots hold soil together and therefore reduce erosion.

Easy areas to increase vegetation

A good figure to aim for is for approximately 30% of a given area of land to be vegetated with trees and bushes. There are various places on a horse property where trees and bushes can be planted where they will not necessarily detract from the available grazing (and may enhance it), and they will be able to carry out their many beneficial functions. As mentioned previously, well positioned trees and bushes may actually increase the yield from your property.

You should research which trees and bushes are best in your area in terms of fire resistance and check that they are not designated weeds or classified as being poisonous to horses. Whenever possible, plant trees and bushes that are native to the area as these will be best adapted to local conditions, soil etc.

The corners of paddocks

The corners of paddocks are dangerous if horses are kept in herds, because they can run into a corner while playing and get trapped. In terms of land management, they are difficult to manage because harrowing them (spreading manure) is difficult. For example, when driving around a paddock with pasture harrows you have to stop, get off, and kick the manure around because the harrow cannot get into the corners. Corners often end up bare, compacted and full of manure and are therefore less productive than other areas of a paddock.

A good solution is to fence the corners off with a simple electric tape/braid - or a more permanent fence. Then, a few trees and bushes can be planted in the corner. The situation is now turned around completely; previously unproductive and potentially dangerous corners will become wildlife havens, providing all the benefits, *and* provide extra shade for your horses etc. – for the cost of a bit of electric tape or braid and a few young plants.

The perimeter of a property

There should be a double fence around the perimeter of a horse property for various reasons:

- A double fence means that you have two fences rather than one keeping your horses on the property.

- A 'living fence' of trees and bushes, backed up by more 'permanent' fencing materials such as a simple plain wire, presents a good solid visual barrier to horses moving at speed and is therefore safer. In fact, whereas horses will attempt to jump an ordinary fence, they will never attempt to jump a 'living fence' if it is high enough.

- A double fence prevents your horses from playing/socialising with a neighbour's horses over a fence. As we have seen before, horses should never be allowed to interact over a fence; beside the physical dangers (fence injuries are high on the list of the cause of death or permanent injury to horses), there are the biosecurity issues when horses on different properties interact.

- If you live in or near suburbia it makes it more difficult for well-meaning people to feed your horses over the fence. This includes people dropping lawn mower clippings and other garden waste in with your horses.

- A double fence means that you can turn a potentially unsafe fence into a much safer fence. A double fence does not necessarily mean double the expense. For example if the property has an undesirable fence already in place around the perimeter, it will probably cost less to plant trees and bushes on the inside of that and then place a simple electric fence on the inside of that, than to re-fence with a new solid fence (which will still only be a single fence).

A row of trees planted around the perimeter of you property provides many advantages.

Between paddocks

Trees and bushes can be planted between paddocks within the property. If you have more than one herd for example, this double fence will allow you to use adjacent paddocks.

As an addition to a shelter

Trees and bushes can be planted so that they enhance man-made shelters or, in some cases, provide total shade for surfaced holding yards. They need to be planted outside the surfaced holding yard so that they are protected from horses by the surfaced holding yard fence. A solid roof can be enhanced with the addition of trees and bushes which will slow the wind without stopping it entirely.

In laneways/driveways

Trees and bushes can be planted in any laneways - if you have to have them in order for horses to take themselves to paddocks (see Appendix: *The Equicentral System*) in which case they may need the protection of a double fence. This second fence can be a simple electric fence. If horses are being led to paddocks, then a second fence will not usually be necessary. The driveway of a property is another example of a good place to plant trees and bushes.

Trees and bushes can be planted between paddocks within the property.

Along contour lines to help channel water

Trees planted in the correct areas can help to move water in a more desirable direction. For example, by planting a row of trees across a hillside and banking the soil a little, water that previously ran straight down the hill can now be channelled sideways and slowed down. Some of the water can be allowed to pass through the line of trees if necessary, which means that instead of the water all travelling down the same path, it is now spread out and benefits a much larger area, in addition to reducing the damage that water travelling in a single channel can cause.

188

In wet areas of land

Water-loving trees can help to soak up excess water, which often occurs in areas that were originally very wet areas e.g. wetlands. By planting the correct trees, you can redress this issue without going to the trouble and expense of installing drainage. Aim for local species whenever possible as these will be better adapted to your conditions.

Trees and bushes can be planted around waterways, high ground, eroded areas, in fact anywhere unsuitable for grazing.

Around waterways

Any waterway that is on or borders a property, whether natural or manmade (such as a farm dam), should have a riparian zone around it.

On steep sections within paddocks

Steeper land is difficult to manage as pasture and therefore is ideal for growing larger vegetation instead, see the section **Revegetation of steeper land**.

Buying and planting

Ensure the plants you buy are healthy specimens. Tube stock which are bushes or trees in narrow tubes, are usually much cheaper to buy and they have a high success rate if they are purchased at the right time (before they outgrow the tube). Tubes allow the roots to grow in the correct way for plant development. If you are revegetating a large area, this is a very good way of buying in bulk. Plan all your purchases well in advance as stocks may be low when you need to plant and you don't want to miss the ideal time to plant by waiting for an order to arrive.

Once you have decided which trees and bushes you want, you will need to find a good supplier. If you are vegetating or revegetating the property bit by bit, it may be possible to get what you need on an ad hoc basis, looking for good buys at nurseries and markets and even auctions. If you are planning to buy in bulk, you may find that your local authority (or other government body) has a nursery. Your local land/soil conservation group is another excellent contact. It may even be possible to get subsidised or free trees when you are vegetating or revegetating a property, particularly if it involves a waterway.

Make sure the trees and bushes that you buy are not poisonous to horses (see the section *Poisonous trees and plants*).

Plant new trees in areas which are inaccessible to the horses; areas such as those described in the section *Easy areas to increase vegetation*. By planting them in these areas, you will find that it is easy to achieve a state where 30% of the property is vegetated with trees and plants (other than grass).

Before you plant, you need to prepare the area. Compacted soil may need to be ripped by a tractor with a ripper so that young roots can infiltrate the soil. Plants grown in waterlogged soil may need mounds around them to prevent drowning. Weeds will need to be controlled before new plants are planted and once your new plants are in the ground, weeds can be further controlled by mulching. Aim to keep a 1m radius clear of grass and other vegetation around your new plant for the first year at least. Aim to plant at the best time of year for your locality. Before you spend money and time preparing the land, buying and planting new vegetation, make sure you speak to local experts, otherwise you may find that your new plantings fail an expensive disappointment.

Protecting vegetation from horses

Trees and bushes need protection if they are to survive and thrive. Even 'fodder trees' need a break and will die out if the browsing of them is not controlled.

The various forms of abuse that horses and other grazing animals can inflict on vegetation include:

- Compaction of the roots by standing at the base of the tree (e.g. using it for shade).

- Chewing the bark and eventually 'ring barking' a tree (where all of the bark is removed right around the trunk). This leads to the death of the tree.

- Eating/trampling young trees and bushes entirely. The plant may not recover, depending on the species.

- Using trees and bushes as a rubbing post. This can cause the tree/plant to snap and possibly die.

Trees and bushes need protection if they are to survive and thrive. Even 'fodder trees' need a break and will die out if the browsing of them is not controlled.

Tree protection

There are various forms of protection for trees and bushes including fencing, wrapping, mulching and using tree savers. What you need will vary depending on the situation of the trees and bushes.

Single trees and copses of trees within paddocks give a property a 'park like' look. Vegetation in the middle of paddock provides shade at all times of the day and if you have very large paddocks, then a few copses of trees *may* be of benefit. Grouped trees tend to be healthier than single trees, as trees rely on each other in various ways. Any gaps between the trees can be allowed to revegetate or be planted with species that you particularly want on the property, such as bird attracting trees.

However, unless they are already established it is better to avoid new plantings within a paddock:

- Trees within a paddock make it harder to work in the paddock (if harrowing/mowing a paddock for example).

- If you are planning on having **The Equicentral System** (see Appendix: *The Equicentral System*) in place, you will want to encourage the horses to use the large shelter rather than the trees.

- Fencing is very expensive and, in order to be effective, it should be out as far as the drip line (the circumference canopy).

If they already exist, then by all means nurture them by protecting them from horses. If you are using electric fencing, this can be more complicated because the electricity will need to be taken underground, out to the area within the paddock, via an insulated carrier.

—

Concentrate on developing any new plantations in the areas described in the section *Easy areas to increase vegetation*, rather than in the middle of paddocks, with an exception being any steep sections of land within a paddock (see the section *Revegetation of steeper land*).

—

Wrapping

If unfenced trees are situated in paddocks, and you are not able to fence them, you need to watch carefully for signs of chewing/ring barking. If the horses start to chew the trees, **there are several things you should think about:**

- Check that there is enough suitable pasture in the paddock. The main reason that horses chew trees is because their diet is not high enough in fibre.

- This can be because there is not enough pasture in the paddock due to over grazing/over stocking, drought etc. In this case the horses need supplementary feed, hay in particular.

- Horses can also crave fibre when pasture is growing rapidly. In this case, it actually has a very high water content – as much as 80% to 90%. Again, the horses need supplementing with hay.

- In both of the above scenarios it is best to remove the horses for all or part of each day and feed hay in surfaced holding yards.

If the trees cannot be protected with fencing, you may need to wrap the trunk/s in mesh, tin or corrugated iron. Keep in mind though that this practice may impact on any local wildlife that climbs trees.

Mulching

Mulch placed around the base of a plant (but not touching the stem or trunk) will help to regulate temperature extremes. Mulch should not touch the base of the plant, as this can cause the stem to rot. Mulch suppresses weeds, acts as a slow release fertiliser for plants (depending on the mulch type), provides an environment for plant-friendly insects and reduces evaporation. Be generous with the mulch and pile it higher on the outer than the inner rim.

Various materials can be used as mulch such as grass (lawn) clippings (although not in areas where horses have access to them, as lawn mower clippings can be dangerous for horses to eat, see the section *A word about lawn mower clippings*), paper, straw/hay or woodchips. In fact, any organic matter will make mulch, with some being better than others. Grass (lawn) clippings are best mixed with coarser material such as twigs and leaves as they tend to clump. Fresh manure should not be used as it can be too 'strong'. Also, when it dries, it can form a surface that repels water. Composted manure should be regarded as a fertiliser and covered with mulch so that it does not dry out. Newspaper, cardboard or carpet placed under another form of mulch will further reduce evaporation by approximately 70%.

A common form of mulch is leaf or bark mulch. Tree prunings are often readily available from businesses that advertise tree lopping/tree care as they are a by-product. Your local authority may have mulch to either sell cheaply or give away (it depends on your local authority). Your local refuse tip may also sell/give away mulch that is made from green waste.

If mulch contains garden waste, make sure your horses do not have access to it as it may contain poisonous plant material.

Any trees that horses stand underneath on a regular basis may need to be mulched to protect the roots from compaction.

Tree savers

Bush and woodland can be sensitive and must be fenced off from domestic grazing animals to protect it. These areas usually provide little or no grazing anyway, so there is often no point in using them for paddocks. Even if they do have good pasture in them, fencing them off will allow you to control the amount of grazing pressure that they receive. If they are left as part of a paddock they will always be available to the grazing animals and will tend to be overused.

Horses like to stand under trees for shelter. This can cause compaction and damage the root system.

Young trees and bushes need to be protected from rabbits, native animals, horses and other grazing animals. Putting a plastic or cardboard 'tree saver' around each plant will protect the plant from small animals such as rabbits and the tree saver also creates a mini greenhouse around the plant, giving it added protection. Tree savers will not protect plants from horses; they will pull the plant out from the top and either eat it, or simply drop it on the ground. Only fencing will protect young trees from larger grazing animals.

Existing vegetation

Areas of established bushland/woodland on the property need to be protected from horses. These areas are often seen as a nuisance on a horse property and are earmarked for clearance (legal or illegal). Instead of being regarded as a problem, they should be seen as a valuable asset for all of the reasons (and more) outlined above. Remember, biodiversity is what you should be aiming for. If not earmarked for clearance, these areas tend to be seen as a handy area for horses to use as shelter. However, unless the grazing and browsing of these areas is controlled, the problems outlined above may occur.

Any fencing should be constructed so that native animals can still pass through. If possible, these areas should link with other tall vegetation areas both on the property and outside the property e.g. neighbours, forests, conservation areas etc. to maintain 'wildlife corridors' (areas that wildlife can move through freely and safely). Leave fallen trees in these areas if safe to do so, as they provide good habitat for wildlife. Your local environmental protection group should be able to help with advice about these areas and, in some localities, financial help may also be available.

Any weeds in these areas must be tackled promptly. These areas tend to get forgotten about when it comes to weed control because they are usually unproductive in terms of proving pasture. As well as causing problems for wildlife and wild (native) plants, noxious weeds will also infest the pastures, so it is far better to curb them before they get out of hand.

Poisonous trees and plants

It is not always easy to identify which plants are harmful to horses. There are no set characteristics that tell us which plants are poisonous, and most of the information about horses and poisonous plants is anecdotal. Fortunately, horses are selective when grazing. This is partly because they are not able to vomit; the entrance valve to a horse's stomach is very strong. This prevents the stomach from rejecting bad food, unlike a dog's stomach, for example, which can and does reject food easily. Horses innately avoid eating plants that they are not sure about and therefore, cases of plant poisoning are reasonably rare.

Possibly the largest factor that leads to domestic horses eating poisonous plants is poor pasture management; if a horse does not have enough of the right kinds of plants to choose from, they will turn to eating plants that they would normally avoid. This is because horses are meant to have fibre available to them at all times. Their digestive system has evolved to deal with large amounts of low energy fibrous food, rather than small amounts of high-energy food, such as what a dog or human has evolved to eat. In turn, their digestive system produces acid on a continuous basis, again in contrast to the digestive system of a dog or human that produces acid as a meal is ingested. As a consequence, when a horse is not able to get enough of the right kind of fibre to eat, they will eat plants that they would normally avoid, otherwise their stomach becomes too acidic and initially becomes painful and eventually the horse develops stomach ulcers.

So, on the whole, horses are good at avoiding plants that are poisonous. They would not have survived to the present day if they did not have some 'knowledge' about what to avoid. As ever though, when horses are kept in a domestic situation, you have to be aware that there are certain circumstances that can change their behaviour and even that of the plants (e.g. 'stressed' plants can become toxic as part of their natural defence mechanism).

Make sure any new or existing plants on a horse property are not poisonous. There are various general factors to remember:

- Some poisonous plants are more dangerous than others - some have to be eaten in large quantities before affecting a horse, while others are poisonous in very small amounts.

- Some are more likely to be eaten than others.

- Some are only poisonous at certain times of the year/or in certain climatic conditions e.g. certain plants that have been stressed by overgrazing and/or drought can become poisonous.

Many plants are poisonous to horses. Bracken is more poisonous at certain times of the year.

There are also several factors about horses that further complicate the issue such as:

- Individual horses differ in whether they will eat poisonous plants or not; this can be affected by how hungry the horse is. A hungry horse will eat plants that they 'know' they should not.

- A hungry horse is also likely to be poisoned more quickly, because the toxins will be absorbed more rapidly.

- Individual horses learn what to eat or avoid when very young by observing their dam. This means that, to some extent, horses that have never been exposed to poisonous plants when young may not know what to avoid when older.

The following precautions must always be observed to minimise risk:

- Learn to recognise poisonous plants in your locality and be suspicious of unfamiliar plants to which horses might have access.

- Find out how such plants can be managed/removed.

- Recognise the conditions when certain plants are at their most dangerous.

- Do not allow horses to have access to garden waste or poisonous trees near fence lines.

- Seek veterinary advice as soon as poisoning is suspected.

Horses suffering from plant poisoning may show a wide range of symptoms including:

- Sudden death or death within a few days.

- Nervous diseases - for example 'Staggers', 'Australian Stringhalt'.

- Photosensitisation.

- Liver damage.

- Kidney damage.

- Skin disorders.

- Heart and lung disorders.

- Intestinal disorders.

When a horse's condition indicates that plant poisoning may have occurred it is important to note:

- What signs of illness the horse is showing.

- What plants the horse has had access to and whether the horse can reach plants over a fence.

- Whether these plants and their toxic properties can be identified.

- Whether neighbours or passers-by may have fed the horses or dumped anything over the fence; this is a common occurrence as inexperienced people do not realise that horses can be susceptible to plant poisoning and can feed horses thinking they are doing them a favour, see the section *A word about lawn mower clippings*.

If poisoning is suspected, seek professional veterinary help immediately and provide the veterinarian with as many answers as possible to the above questions.

Prevention is better than cure, so horse owners should be aware of the possibility of poisoning and at least be able to identify the most common poisonous plants both in their local area and in general so that any new noxious weeds springing up in an area can be quickly identified and taken care of. Always be aware of any new or unusual plants appearing on the property. Find out if your local authority has a weeds officer and buy or borrow a good, well-illustrated book that will help you to identify weeds and poisonous plants. The Internet is an excellent resource for identifying and finding out more about poisonous plants.

—

Get local advice about what kind of trees to plant near horse facilities. Considerations include weed status, fire resistance, likelihood of dropping branches, safety in terms of poisoning and which animals they will attract.

—

On the whole, horses are good at avoiding plants that are poisonous, but poor pasture management will add to the risk of them eating the wrong plants.

A word about lawn mower clippings

Never feed a horse on lawn mower clippings or any other types of garden waste as this can be a very dangerous practice for various reasons:

- When lawn mower clippings are freshly cut (and for a period of time afterwards) they are fermenting - this is why a pile of lawn mower clippings are warm or even hot to the touch.

- When a horse is given a pile of fresh lawn mower clippings to eat, they tend to gorge on them. This is because the clippings have been chopped up into small pieces (by the mower) so they do not need to be chewed.

- The horse can simply swallow them quickly, without mixing them with saliva (which is what normally happens when a horse chews their food).

- Therefore, the lawn mower clippings arrive in the stomach already fermenting and without the benefit of saliva to 'dilute' them.

In the normal situation, grasses that are eaten by a horse do not start to properly ferment until they are further along the digestive system, in a part that has specialised microorganisms to help the fermentation process.

The fermenting gases given off by the lawn mower clippings can expand to the point where they rupture the stomach (which is fatal!). Even if this does not happen, the fermenting clippings can cause colic or other very dangerous gastrointestinal disorders as they move through the rest of the digestive system.

The lawn mower clippings may also contain the clippings of garden plants that the horse would normally avoid. You cannot see them and your horse will probably not be able to detect them either. Horses are usually quite good at avoiding eating poisonous plants but if they are mixed in with grass clippings they may eat them inadvertently.

The lawn that the clippings came from may have been sprayed with chemicals (such as fertilisers, pesticides and/or herbicides) before it was mown. The horse will then ingest these chemicals along with the grass clippings.

If your horses live on land that is alongside one or more suburban properties, be aware that the neighbours may be dumping garden waste including lawn mower clippings over the fence, therefore it can be a good idea to erect a double fence in this situation. Even a simple electric fence that is placed several feet to the inside of the perimeter fence may be enough, although neighbours may still insist on putting garden waste in with your animals. Remember that in or near suburbia, you may be required by law to erect signage warning people that electric fencing is being used - check with your local authority.

Other measures that you can take include posting notices asking people not to feed your animals. Chat to neighbours about the risks and ask them to stop dropping any garden waste over the fence. In most cases, people think they are doing your animals a favour and, once the dangers are explained to them, they will stop doing it.

If you are a keen composter of your manure, you may be able to make good use of their garden waste on your compost pile. For example, if you have lots of woody products such as wood shavings to compost, the addition of green garden waste will be a boon so you could ask them to put the garden waste on your manure heap instead.

What about when a paddock is mowed in order to tidy it up? The horses may be able to be left in a paddock after mowing provided that the cuttings are well scattered (they dry out quickly in this case). The mown grass cuttings in this situation are not usually cut as fine as are lawn mower clippings and they usually dry out quickly or decompose with little or no fermentation. The remaining freshly cut grass stems (rather than the cuttings) are sweetened with sugars stored in the plant leaf, so the grazing animals are usually more interested in this rather than the grass cuttings themselves. Be aware though that horses that are at risk of laminitis will be at risk of eating too much sugar in this case.

If you are at all unsure, remove the horses from the paddock for several days before letting them back in. Better still, only mow paddocks *after* you move the herd to the next paddock as part of a rotational grazing management system.

Never feed a horse on lawn mower clippings or any other types of garden waste as this can be very dangerous.

The following section/s are directly reproduced from *The Equicentral System Series Book 1 – Horse Ownership Responsible Sustainable Ethical* in order to aid your understanding in this book.

Appendix: The Equicentral System

The Equicentral System is a *total* horse and land management system that we have developed and have been teaching to horse owners around the world for many years now. It uses the natural and domestic behaviour of horses, combined with good land management practices, to create a healthy and sustainable environment for your horses, the land that they live on *and* the wider environment.

There are many examples in various countries including Australia, New Zealand, the UK, the USA and even Panama! This list just keeps growing as people realise the huge benefits of using **The Equicentral System** in order to manage their horses *and* their land in a sustainable way.

How The Equicentral System works

The Equicentral System utilises the natural grazing and domestic paddock behaviour of horses in order to benefit the land that they live on *and* the wider environment. In turn this benefits the horses and it also benefits you (and your family) because it saves you money and time/labour.

- The main facilities – water, shade/shelter, hay and any supplementary feed are positioned in a surfaced holding yard so that the horses **can always** get back to them from the pasture they are currently grazing.

- The watering points are **only situated in the surfaced holding yard**, instead of there being one in each paddock. If you already have water troughs' situated in paddocks these can be turned off when horses are using the paddock, and turned back on again if and when other animals, such as cows or sheep, are grazing there.

- Individual water troughs/drinkers (or buckets) of course are also needed in any individual yards/stables.

- If possible all of the paddocks **are linked** to this surfaced holding yard area, although only one paddock is in use at any time.

- The gate to the paddock that is currently in use **is always** open, so that the horses **can always** get themselves back to the water/shade/shelter etc. **In short, the horses are never shut out of the surfaced holding yard**.

- Occasionally they may be fastened in the surfaced holding yard (with hay), but this is usually for the purpose of preventing damage to the land and increasing healthy pasture production.

- Apart from trees or bushes that are situated in/around paddocks, **the only shade/shelter is in the surfaced holding yard**. This shade/shelter is very important. It should be large enough for the whole herd to benefit from it at the same time.

The Equicentral System: *all of the paddocks lead back to the surfaced holding yard. There is shade/shelter and water in this central area. Hay can also be fed here.*

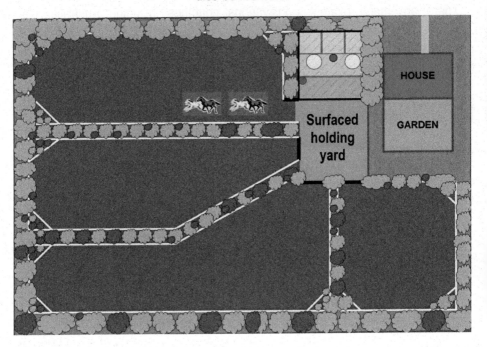

Additional information

- The surfaced holding yard area *can* also be a riding/training surface if that is what you wish. You may prefer to keep it separate or indeed you may not need a riding/training surface - but if you do then this means that the expense of creating this surfaced area has double benefits. This could mean that you are able to afford and justify this surfaced area sooner because you are going to get more use out of it. Also remember, the smaller the property, the more the facilities need to be dual purpose whenever possible so that you have as much land in use as pasture as possible.

- Careful consideration of the surface is required, especially if it is to be a riding/training surface as well. Remember - bare dirt is not an option. Wet mud is dangerously slippery and can harbour viruses and bacteria which can affect the horses legs. Mud will become dusty when dry, meaning that the horses will be breathing in potential contaminants *and* you will be losing top soil.

The surfaced holding yard area can also be a riding/training surface if that is what you wish. If you do then this means that the expense of creating this surfaced area has double benefits.

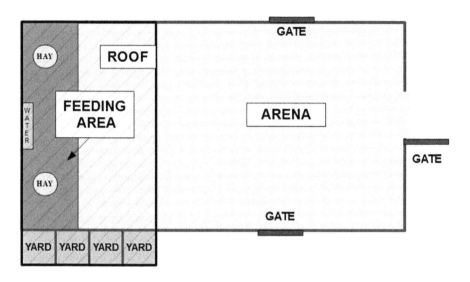

See the third book in this series for more information about using a surfaced holding yard as a riding surface as well, *The Equicentral System Series Book 3 – Horse Property Planning and Development.*

—

- It is useful if there are also some individual holding yards (or stables if you already have them) – preferably linked to the surfaced holding yard for ease of use. You can then separate horses into them for any individual attention that they may require (such as grooming, supplementary feeding etc.) or for tacking up etc. You can also put the surplus horses in them if you are riding/training one of the herd members on the larger surfaced holding yard.

- This system is *not* about food restriction – quite the opposite. It is about transitioning horses to an ad-lib feeding regime of low energy pasture plants and hay (see the section *Changing your horse/s to 'ad-lib' feeding*) so that they no longer gorge and put weight on because of it.

- Your pasture may need to be transitioned to lower energy plants. This does not always mean reseeding.

This is too large a subject to cover here and is covered in the second book in this series *The Equicentral System Series Book 2 - Healthy Land, Healthy Pasture, Healthy Horses*.

—

- Hay can be fed in the larger surfaced holding yard if the horses get on well enough; generally horses will share hay. Otherwise, it can be fed in the individual holding yards/stables, but keep in mind that there should *always* be some form of feed available to the horses.

- It can be a good idea to create a feeding area in the larger surfaced holding yard using large rubber mats or similar.

It is useful if there are also some individual holding yards (or stables if you already have them) – preferably linked to the surfaced holding yard for ease of use. These individual yards can be made from swing away partitions if you only require them periodically (picture left) or can be permanently in place (picture right).

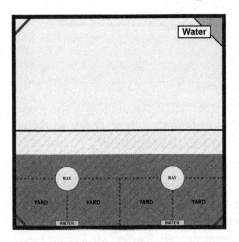 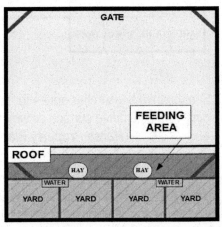

- **The Equicentral System** works best on a property where the horses live together as one herd, otherwise you will need to replicate it for each group of horses that you have. However, many of our clients have done just that in the case of larger properties with various classes of horses (for example studs, livery yards etc.).

- **The Equicentral System** assumes that you already have good grazing management in place (rotational grazing) or that you plan to implement it. Remember, rotational grazing involves moving the animals around the land as a herd, one paddock at a time, rather than allowing them access to the whole property at once (set-stocking).

The Equicentral System in practice

This is an example of how **The Equicentral System** works in practice. In this example, the horses are being kept in the surfaced holding yard at night (or in individual holding yards/stables) and out at pasture through the day, but remember - if there is enough pasture, then the horses do not need to be confined overnight unless you have other reasons for doing so.

- In the morning you open the surfaced holding yard/riding arena gate and the horses **walk themselves** to the paddock that is currently in use for a grazing bout (which lasts between 1.5 to 3 hours), the gate to this paddock should already be open (the other paddocks should have closed gates as they are being rested).

- At all times the horses are free to return to the surfaced holding yard for a drink, but they usually won't bother until they have finished their grazing bout.

- After drinking, the shade and inviting surface in the surfaced holding yard encourages the horses to rest (loaf) in this area before returning to the paddock for another grazing bout later in the day.

After a grazing bout, the horses return to the surfaced holding yard for a drink.

- Leaving hay in the surfaced holding yard can encourage even more time being spent (voluntarily) in the surfaced holding yard and less time spent in the paddock.

- At the end of the day, the horses return from the paddock to the surfaced holding yard to await you and any supplementary feed that they may be receiving.

- You simply close the gate preventing them from returning to the paddock for the night, or, if conditions allow it, the horses can come and go through the night as well as through the day.

After drinking, the shade and inviting surface in the surfaced holding yard encourages the horses to rest (loaf) in this area before returning to the paddock for another grazing bout later in the day.

The Equicentral System benefits

The Equicentral System utilises the natural and domestic behaviour of horses to better manage the land that they live on. **This system of management has many, many benefits including:**

Horse health/welfare benefits:

- It encourages horses to move more and movement is good; a grazing horse is a moving horse. A recent (Australian) study showed that horses in a 0.8 Ha. paddock walked approximately 4.5km per day, even when water was situated in the paddock; additional movement to the water in the surfaced holding yard, therefore, further increases this figure. More movement also means better hoof quality as the hooves rely on movement to function properly. Remember a healthy, biodiverse pasture encourages more movement.

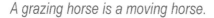

A grazing horse is a moving horse.

- It *maximises* time spent grazing for horses and aims to avoid food restriction. By confining horses initially and when the weather is very dry or very wet (or in order to transition horses that have been on restricted diets in the past), so that the pasture begins to improve, becomes more biodiverse etc. they will be able to graze more in the future because healthy pasture can withstand more grazing.

- Horses are not being forced to stand in mud, especially around gateways when weather conditions are wet. Horses are not good at coping with continuous wet conditions – hence the ease with which they develop skin conditions such as greasy heel/mud fever. Remember - in the naturally-living situation, they will *take themselves* to higher, dryer ground to loaf, even if the areas that they graze are wet. When horses are fastened in wet paddocks, they do not have this choice.

The horses will be able to graze more in the future because healthy pasture can withstand more grazing.

- Eliminating mud also means eliminating dust because they derive from the same thing (bare soil) at either end of a spectrum. Apart from the obvious benefits to not losing your top soil, this means that horses (and humans) do not have to cope with living in a dusty environment.

- Horses move around a paddock in a natural fashion, choosing what to eat. As rotational grazing increases the diversity of plants in a pasture, the horses benefit from access to a larger variety of plants. This means that the horses eat a more natural, varied diet. In addition, healthier plants are safer to graze than stressed, overgrazed plants (more fibre and less sugar per mouthful).

—

See the second book in this series *The Equicentral System Series Book 2 - Healthy Land, Healthy Pasture, Healthy Horses* for much more information on this subject.

—

- The stress of not being able to get to food at will, along with all of its associated problems, is removed. Horses with different dietary requirements can be catered for with the addition of supplementary daily feed (in a separate but preferably adjoining area).

- The surfaced holding yard is seen by the horses as a good place to be and therefore, if and when it is necessary to fasten them in there, (bad weather, for the vet etc.) they are not stressed.

- The horses now have choice; they can choose as a herd whether to graze, walk to the 'water hole', snooze in the shade etc. Instead of us deciding when a grazing bout will start/finish, the horses can decide for themselves. These are all behaviours that naturally-living horses take for granted, but domestic horses are usually 'micro managed' in such a way that a human decides where they will be at any point in the day. This might not seem like a big deal but it really is.

The Equicentral System provides a home range whereby the horses can access the available resources in a more natural fashion.

Time saving benefits:

- You do not have to lead horses in and out to the pasture. The horses are waiting for you close to the house, or at least in an area where you need them to be, much of the time. If they are currently grazing, you simply call them; horses soon learn to come to a call for a reward. This means that when you return from work and the weather is bad, you do not need to trail out in the wind and the rain to bring them in, they will be waiting for you in the surfaced holding yard.

- You do not have to spend time carting feed around the property (or keep a vehicle especially for the job) because the horses *bring themselves* to the surfaced holding yard for feed. The horses *move themselves* around the property, *taking themselves* out to the paddock that they are currently grazing, and bringing *themselves* back for water and feed.

- The single water trough in the surfaced holding yard is all that you have to check each morning and night, saving you having to go out to a paddock and check the water.

- It is far quicker to pick up manure from the surfaced holding yard than from pasture if you collect your manure.

- Any time you save can be spent on other horse pursuits such as exercising them!

Cost saving benefits:

• Money spent on the surfaced holding yard is money well spent, as this area is used *every day of the year* for *at least twelve hours* a day, even if you are not also using this surface for riding/training.

• Money spent on vet bills for treating skin and hoof conditions is reduced or totally eliminated.

• The expense of installing and/or maintaining a water trough in each paddock is spared.

• The expense of installing individual shade/shelters in each paddock is spared. Instead, one large shade/shelter is erected at the side of or over the surfaced holding yard, which means you may end up with a partially covered all weather riding/training surface if you are multi-tasking this area!

• This large shade/shelter will be in use *every day* of the year, unlike shade/shelters that are situated in paddocks and are only in use when the paddock is in use. Remember - if you are rotating your paddocks (as part of a rotational grazing management system), then this means that each paddock will be empty, and any shade/shelters situated in them will be unused, for a large part of each year.

• Annual maintenance including time and expense, of numerous shelters (especially if they are made of wood) is avoided adding to the cost effectiveness of **The Equicentral System**.

• Many horse properties already have the facilities required to implement **The Equicentral System**. Often the required infrastructure either already exists on a horse property, or the property needs minimal changes.

- Laneways (and their associated costs) can be kept to a minimum. In areas that *do* require laneways, any money spent on surfacing them is well utilised as the laneways will be used by the horses several times a day.

This large shade/shelter will be in use every day of the year, unlike shade/shelters that are situated in paddocks and are only in use when the paddock is in use.

- Better land management means more pasture to use for grazing (and safer healthier pasture) and more opportunities for conserving pasture (as 'standing hay' for example) or making hay. This all leads to much less money being spent on bought-in feed.

- You do not need to buy and maintain a vehicle for 'feeding out'. The horses come to where the feed is stored rather than you having to trail around the paddocks 'feeding out'.

- Setting up **The Equicentral System** will not devalue your land. It will actually increase the value of it through good land management. Likewise, if you sell the property, the next owner can choose to set up a more traditional management system by putting water and shade/shelter in every paddock if they wish.

Safety benefits:

- Horses move themselves around the property, therefore there is less unnecessary contact between humans and horses. This is an important point if you have (usually less experienced) family or friends taking care of your horses when you are away. **The Equicentral System** allows them to see to your horses without them having to catch and lead them around the property.

- It reduces or eliminates the incidence of horses and people being together in a paddock gateway. When horses are led out to a paddock, they can be excited because they are about to be freed; and when they are waiting at a gate to come back in for supplementary feed, they are keen to get through the gateway in the other direction for their feed. Horses can crowd each other and human handlers can become trapped. These situations are very high risk on a horse property.

- Depending on its position, the surfaced holding yard can be a firebreak (for your home) and a relatively safe refuge in times of fire/storm/flood for your horses. The layout of the property may result in the horses being pushed (by rising water) back towards the surfaced holding yard in a flood. Assuming the surfaced holding yard is built on higher ground, this can save lives! By training the horses to always come back on a call, you can get them into the surfaced holding yard quickly in any emergency situation. This makes it far easier for you, your neighbours, or the emergency services to evacuate your horses if necessary in emergency situations.

Land/environmental management benefits:

- **The Equicentral System** is a *sustainable* system that acknowledges that a horse is *part of* an ecosystem, not separate to it.

- **The Equicentral System** complements a rotational grazing land management system and allows for the fine tuning of it. Remember - rotational grazing encourages healthy pasture growth and aids biodiversity by moving the animals to the next grazing area before they overgraze some of the less persistent plant varieties.

- With good land management, the productivity of biodiverse, safer pasture should *increase* rather than decrease over time, leading to fewer periods when it is necessary to fasten horses in the surfaced holding yard over time. Remember - biodiversity is good for horses *and* good for the environment.

- It *vastly* reduces land degradation that would be caused by unnecessary grazing pressure. The horses *voluntarily* reduce their time spent on the pasture

214

They will tend to spend the same amount of time grazing (as they would if they were fastened in a paddock for 24 hours), but will tend to carry out any other behaviours in the surfaced holding yard.

They will tend to carry out any other behaviours in the surfaced holding yard.

- They prefer the surfaced holding yard not least because, if it is situated near the house, or at least in an area that they can see you coming towards them, no self-respecting horse will miss an opportunity to keep watch for the possibility of supplementary feed! The water and shade in the surfaced holding yard also encourages the horses to loaf in this area. If the horses are allowed to come and go night and day they will reduce the grazing pressure (grazing pressure being a combination of actually eating but also standing around on the land) by approximately 50%. If you fasten them in the surfaced holding yard (with hay) overnight, you will further reduce the grazing pressure by about another 50% (making a total of about 75%). This reduction in grazing pressure will make a *huge* difference to the land.

- The corresponding compacted soil/muddy areas that surround water troughs and paddock shelters, as well as the tracks that develop in a paddock are avoided. Bare/muddy/dusty gateways are also a thing of the past as horses are *never* fastened in a paddock waiting to come in. Don't forget that the idea is to reduce any unnecessary pressure on your valuable pasture and increase movement. Remember - the reason horses stand in gateways is because that is usually the nearest point to supplementary food; they are either fed in that area or their owner leads them from there to a surfaced holding yard or stable to feed them. If the gate is closed, they stand there; if the gate is open, they bring themselves into the surfaced area, which becomes their favourite loafing area.

—

Managing your land in this way results in less or no soil loss, in fact if you manage your land well you should be able to increase soil production – see the second book in this series *The Equicentral System Series Book 2 - Healthy Land, Healthy Pasture, Healthy Horses*.

—

- It reduces the area of land used for laneways and therefore the land degradation caused by them – by minimising laneways as much as possible. This is done by creating a layout whereby the paddocks lead directly to the surfaced holding area or by creating temporary laneways. If paddocks are already fenced and laneways are in place then this system utilises them efficiently and safely e.g. the horses are not fastened in narrow areas, they can spread out when they reach the paddock at one end or the surfaced holding yard at the other.

- It increases water quality – by minimising or eliminating soil and nutrient runoff. Rotational grazing maintains better plant cover – the absolute best way to keep soil and nutrients on the land and out of the waterways.

- Strip grazing is usually easier to set up because the water point is back in the surfaced holding yard, meaning that the fence only has to funnel the horses back to the gate, without having to take the water trough position into consideration.

- Hay is fed in the surfaced holding yard area rather than the paddocks allowing for better weed control.

Public perception benefits:

- **The Equicentral System** helps to create a positive image of horsekeeping.

- **The Equicentral System** is most likely to be regarded as a good way to manage land by landowners, the general public and the local authorities. There is a general expectation that land should be well managed - e.g. less mud/dust and fewer weeds rather than more mud/dust and weeds.

- **The Equicentral System** fulfils this expectation, creating a positive image of horse ownership rather than a negative one. This is an important point, remember - in some areas legislation is being pushed forward to reduce horsekeeping activities due to the negative image caused by the often poor land management practices on many horse properties.

- In particular, as horses are often kept on land that is leased rather than land that is owned by the horse owner, the landowner usually, and quite rightly, expects to see good land management taking place. Of course horse owners that own their own land should, and usually do, want the same.

There is a general expectation that land should be well managed - e.g. less mud/dust and fewer weeds.

Manure and parasitic worm management benefits:

- The manure, along with the horses, comes to you. More manure is dropped in the surfaced holding yard and much less in the paddocks (as much as 75% less if you fasten the horses in the surfaced holding yard/s at night with hay). This allows for much better manure management.

- If you usually collect manure that is dropped on pasture then it is physically easier to pick up manure from the surfaced holding yard/s.

- This collected manure can then be composted (which also reduces parasites on your property, as thorough composting can kill parasitic worm eggs and larvae).

- Composted manure is much better 'product' than 'fresh' manure.

- Less manure on the pasture is less importunity for parasitic worm larvae to attach to pasture plants.

- Better manure management also means less reliance on worming chemicals.

- The extra pasture created by managing the land better increases the possibility of being able to 'cross-graze' (graze other species of animals on the land). This further reduces parasitic worms on the land in the most natural way possible, because parasitic worms are what is termed 'host specific', meaning that they can only survive when picked up by the host animal that they evolved alongside.

- Rotational grazing also aids in parasitic worm management by increasing the time that a given area of pasture is resting, which means that *some* of the parasitic worm larvae (on the pasture) dry out and die as they wait - in vain! - for a horse to eat the plant that they are attached to.

—

See the second book in this series *The Equicentral System Series Book 2 - Healthy Land, Healthy Pasture, Healthy Horses* for detailed information about manure management including many novel ideas such as how chickens can be used to help you to manage horse manure.

—

Manure dropped on the surfaced yard rather than pasture is also far preferable in terms of parasitic worm management (no plants for any larvae that hatch out to attach to).

Implementing The Equicentral System

This section describes some of the practicalities of implementing **The Equicentral System**. In some sections you will be referred on to one of the other books in the series because there is too much information to be covered here.

On your own land

Obviously, this is what most people aspire to; having their own land. If you are in this fortunate position, then you are free to set up **The Equicentral System** and reap the benefits.

If you are fortunate enough to own your own land then you are free to set up **The Equicentral System** *and reap the benefits.*

On small areas of land

The Equicentral System is the ideal way to manage horses when there is only a small area of land available for grazing. In fact, it is the only way that will ensure that the horses have grazing available to them and at the same time, land degradation is not created. Of course it will mean that the horses are not able to graze as much as they or you would like, but at least the grazing they do have will be 'quality' grazing rather than standing around on bare, dusty/muddy, weedy land. So don't ever think that your situation would not support **The Equicentral System**, because it will.

At least the grazing they do have will be 'quality' grazing.

Source - Alayne Blickle of Horses for Clean Water - USA.

A common scenario, especially when people lease land, is that they have just one paddock (picture A). Even in this situation it is not difficult to set up **The Equicentral System**. You can still make huge changes to your management of the land and horses by creating a fenced hard standing area by the gate (surfaced holding yard) preferable with a shade/shelter. The 'paddocks' can fan out from this area (picture B). The facilities/fencing can be made temporary/relocatable materials including sectional holding yard fences and rubber paving mesh for the surfaced holding yard and electric fencing for the internal fences.

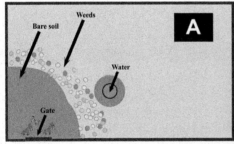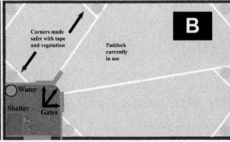

On large areas of land

The Equicentral System works well on a large 'mixed use' property as well as on a large horse property such as a stud. Other species of grazing animals respond well to having a centralised area for resources so it is possible to set up multiple central holding areas on large properties that have different types of grazing animals. Likewise, a large horse stud that has various age groups of horses can also have multiple central points.

In this example (on a 100 ac/40 ha property) there is an **Equicentral System** *set up for the horses, positioned near the house as the owners will be handling the horses much more regularly than they will the cattle. The central area for the cattle is at the bottom of the hill, well away from the house. Occasionally the owners can bring the cattle up the hill (via the driveway or through various paddocks) so that the horse paddocks can benefit from cross grazing.*

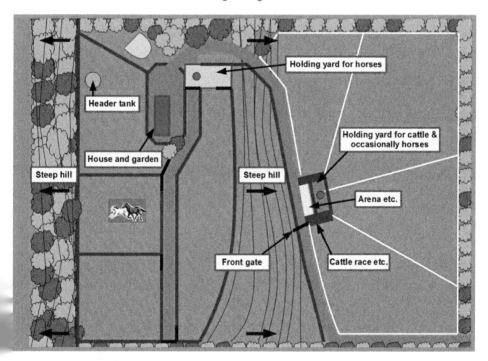

Two **Equicentral Systems** *in place. When the youngstock need to be brought up to the main yard near the house, the mares and foals are fastened in one of the paddocks.*

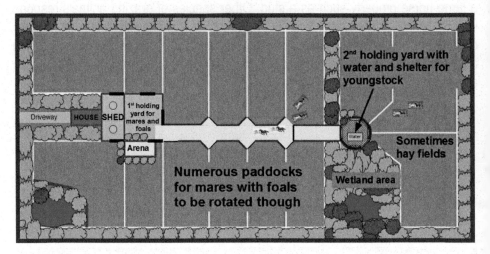

In different climates

The Equicentral System works equally well in any climate, whether it be temperate/wet/tropical/dry/arctic etc. This is because you are providing a home range for the horses and allowing them (most of the time) to make the decisions about when to graze and when to shade/shelter. So, when insects are particularly problematic, they can *take themselves* to the shade to escape them and when they need fibre, they can decide for themselves when to go out to graze or to stay in the shade/shelter and eat hay (if you leave 'ad-lib' hay in the shelter). When it is very cold and wet, they can decide to shelter mainly at night and graze mainly by day. You only need to step in and 'micro manage' them when they are about to put too much pressure on the land that would lead to less pasture in the future.

Yes it is good to understand different land class types and understand what sort of soil you have, but initially it is more important that you understand that land and climates tend to range from too wet to too dry.

Either way, as long as you have an area to allow the horses to remove the pressure voluntarily, as well as involuntarily when you decide that the land needs a hand; you will see a reduction in, and eventually an elimination of dust and mud and its associated problems.

Using existing facilities

If your land already has facilities in place, **The Equicentral System** can usually be implemented without making any major structural changes to your property. **For example:**

- Hard standing that is already in place around any farm buildings/stables etc. should be able to be utilised as a surfaced holding yard. So, if you already have a 'stable yard' that has hard-standing with the paddocks leading out from this area, then you already have a great set up.

- In many cases it is just a matter of leaving the gate to the paddock that is currently in-use open so that the horses can get back to this area, rather than fastening them on the other side of the gate.

- Old farm buildings can usually be used to great effect, as long as they are safe and have a high enough roof for horses. Such buildings often already have hard standing in and around them.

An old farm building such as this would be great for converting to a 'run-in shed' for horses.

- By implementing rotational grazing and always having the gate open to the paddock they are currently using, the horses will bring themselves back to the yard and stand on the surfaced area, rather than stand in the gateway and create mud.

- You can turn off the water in the paddocks (or stop carrying water out to the paddocks!), and set up a water trough on the hard-standing area.

- You may want to create extra shade/shelter for the horses by using the existing buildings to fasten 'shade sails' from, or extend the roof area with a more solid style of roof.

- If you have a block of stables you may decide to open the fronts of some or all of the individual stables boxes to create a 'run-in shed'. Keep in mind that, for various reasons such as tacking up, health care/vet work/trimming/shoeing, supplementary feeding etc., it is still useful to have some individual holding areas.

- Surplus stables can be used for storing hay etc.

By implementing rotational grazing and always having the gate open to the paddock they are currently using, the horses will bring themselves back to the stable yard and stand on the surfaced area, rather than stand in the gateway and create mud.

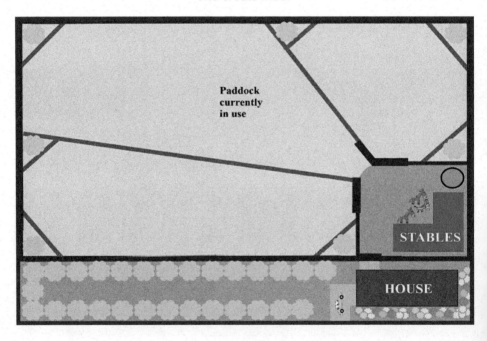

You may want to create extra shade/shelter for the horses by using the existing buildings to fasten 'shade sails' from, or extend the roof area with a more solid style of roof.

Source - Alayne Blickle of Horses for Clean Water - USA.

On land that you lease

A landowner should be happy for you to implement a system of management that is going improve their land value. For example, horse owners often lease land from farmers and most farmers understand the value of a rotational grazing system.

If you need more land and you are already demonstrating good land management techniques, then you are more likely to be given the opportunity to lease/use that additional land than someone who is not doing so. For example, it is not uncommon for neighbours that have land they are not using to offer it to someone with grazing animals. However, they are unlikely to do this if they see that the land you are currently using is badly managed.

In this situation you may want to use facilities/materials that can be removed and taken with you if you ever move on. There are various options for temporary/relocatable shade/shelters (that have the added advantage of not usually requiring planning permission), fences (including sectional holding yard fences) and even surfaces (such as rubber paving mesh).

—

See the third book in this series *The Equicentral System Series Book 3 – Horse Property Planning and Development* for lots of ideas and solutions.

—

If you need more land and you are already demonstrating good land management techniques, then you are more likely to be given the opportunity to lease/use that additional land.

On a livery yard (boarding/agistment facility)

It is perfectly possible to have this system in operation on a horse livery yard. If the horses already live in herds, then the issues are just the same as for setting this system up on a private-use property. If they do not, then first you have to establish the logistics of how you will integrate the horses into herds. There are several options. You may decide to have a mare group and a gelding group (or several depending on the numbers). You may have a variety of groups, for example, you may decide to let owners group their horses so that they are able to share horsekeeping duties with friends.

Hopefully, the property does not currently have a single horse shade/shelter in every paddock as these will be too small for grouped horses.

Small paddocks that previously housed one horse each can now be rotationally grazed. A central holding yard area will need to be constructed for each herd.

If you wish to implement this system and you would prefer that the owners do not enter a paddock containing a large herd of horses, then you can create a routine whereby the horses come into individual areas once or even twice a day (preferably these areas should lead off from the large surfaced holding yard). This can be very useful in cases of horses receiving different levels of supplementary feed etc.

With single horses in 'private paddocks'

Please note: we are not advocating keeping horses separate to each other, but some owners will *never* put their horse with another horse and very occasionally there are good reasons for separating horses.

Separated horses can still benefit from better pasture management and a better shade/shelter arrangement that allows some socialisation, if it is not already in place.

Horses in 'private paddocks' should have access to a shade/shelter, which should be positioned at the gateway (within a surfaced holding yard) and alongside the next 'private paddock', so that two horses can socialise.

Horses in 'private paddocks' should have access to a shade/shelter, which should be positioned at the gateway (within a surfaced holding yard) and alongside the next 'private paddock', so that two horses can socialise.

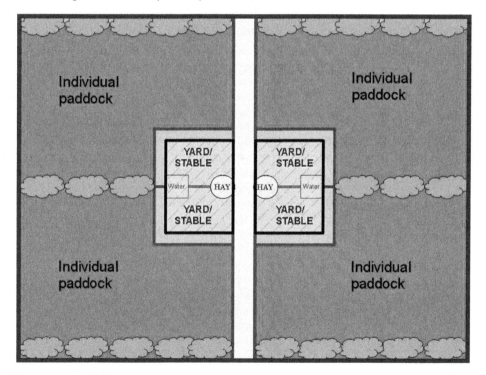

We do not advocate horses socialising over fences, but in this situation, if the partition between the two horses is solid from their chest down (so that they cannot injure a leg) and open above chest height (preferably totally open rather than 'caged'), then two separated horses can and will happily spend many hours 'hanging out' in this area rather than on the sensitive pasture.

When the weather is too wet or too dry the horses can be temporarily prevented from going out of the yard and causing damage that will lead to less quality grazing in the future.

By subdividing any pasture that is available to such horses (as long as this does not create ridiculously small areas) and rotating them around these areas, the land has time to rest and recuperate, resulting in better quality grazing for the horses into the future. For the pasture, any rest is better than none.

See the section *Temporary laneways* for ideas about subdividing smaller/awkward areas of land.

Starting from scratch

If you are in the fortunate position of being able to start a horse facility from scratch, you have much planning to do.

Setting up **The Equicentral System** from scratch should cost less than setting up conventional facilities. Money that would be spent on items such as stables, individual shelters etc. can instead be spent on a surfaced holding yard/s with shelters. Don't forget you can create a surface that can be used for riding/training as well!

—

See the third book in this series *The Equicentral System Series Book 3 – Horse Property Planning and Development* for more information about using a surfaced holding yard as a riding surface, and for information about the total planning of a horse property.

—

Minimising laneways

Depending on the layout of the property, it should be possible to minimise laneway usage, particularly if the internal fences are not yet established.

If you are utilising **The Equicentral System** the horses will be living as a herd, moving *themselves* around the property, only ever having access to one paddock at a time, therefore the paddocks should be arranged so that they lead either directly back to the surfaced holding yard or, if this is not possible (in the case of a long narrow property for example), temporary laneways can be constructed (see the section *Temporary laneways*).

Aim to *minimise* laneways for several reasons:

- Laneways take up space that could otherwise be used as pasture for grazing.

- Laneways concentrate hoof activity to a narrow strip and therefore create land degradation problems such as mud/dust, soil erosion, weeds.

- Laneways are difficult to harrow, mow, weed etc.

- Laneways require more fencing and sometimes require surfacing, therefore extra expense.

Laneways take up space that could otherwise be used as pasture for grazing.

If possible, create a layout for your land that reduces or eliminates the need for laneways. There are several ways that you can do this:

- You may be able to have your paddocks 'fan-out' from the surfaced holding yard so that all paddocks lead directly back to this area without the need for laneways.

- You can utilise 'temporary' laneways so that the land used as a laneway is only used as such when necessary and becomes part of the paddock again when not needed as a laneway (see the section **Temporary laneways**).

   Property A has a laneway leading to the far paddock but if possible, lay the property out so that there are minimal or no laneways (property B).

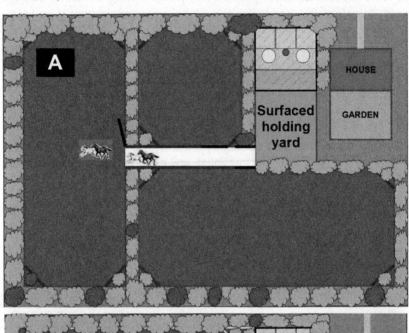

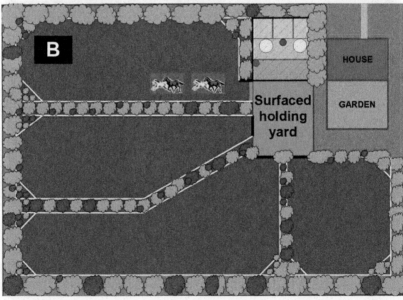

Temporary laneways

A *temporary* laneway can be constructed from temporary electric fence posts (sometimes called 'tread-ins') and electric fence tape to create a 'laneway' that takes the horses to the far end of a long narrow paddock, or even across one paddock to a paddock beyond. This is sometimes preferable to erecting a permanent laneway, because it can be *removed* when the horses are grazing the near section or the near paddock. The land that *would* become a permanent laneway is spared and can be managed as part of the paddock for some of the year; it is far easier to manage land as part of a paddock than to manage land as part of a laneway.

An alternative to using temporary electric fence posts is to put permanent fence posts (e.g. pine poles or steel posts with plastic caps on) in a line where you need them and fasten electric tape carriers to them. This way, an electric tape can be run out through them when necessary and reeled back in when it is not needed.

—

See the third book in this series *The Equicentral System Series Book 3 – Horse Property Planning and Development* for information about fences, including electric fences.

—

A temporary laneway can be used to take horses to a far paddock on a narrow property.

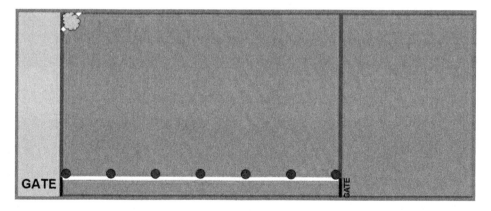

To strip graze a paddock using a temporary laneway, the first stage would look like this....

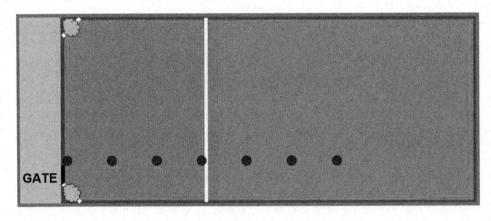

...the second stage would look like this....

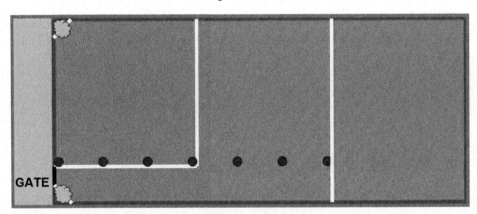

...and the third stage would look like this....

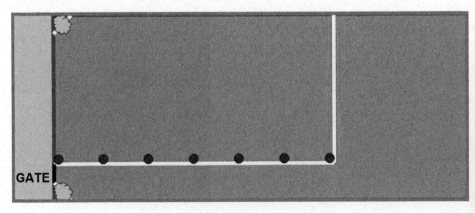

Constructing a holding area

It is imperative that you have a surfaced area for horses to stand. Otherwise, you are quickly going to have mud and dust, soil loss, weeds etc. You are also going to see the skin conditions that are associated with mud such as mud fever/greasy heel etc. If you need to construct a purpose built holding area (rather than utilise something that is already in place) you will need information about this subject.

—

See the third book in this series *The Equicentral System Series Book 3 – Horse Property Planning and Development* for lots more information about constructing a surfaced holding yard.

—

Constructing a shade/shelter

It is imperative that you have shade/shelter for your horse/s. This will increase their need to move themselves back to the holding area and is important for protection both from inclement weather and from insects. There are a huge variety of options ranging from traditional to non-traditional, and from permanent to temporary/relocatable.

—

See the third book in this series *The Equicentral System Series Book 3 – Horse Property Planning and Development* for lots more information about shade/shelters.

—

Fencing considerations

As a general rule, your external (perimeter) fence and any areas in which horses are being confined in a smaller space should have good solid permanent fencing. Anywhere that horses can move away from each other can, if necessary, be fenced inexpensively with electric fencing, certainly initially. Avoid having electric fencing around the holding yard if possible and, if you do use it in laneways, be aware that horses can knock each other into it and it can therefore be stressful for horses when they cannot get out of each other's way.

—

See the third book in this series *The Equicentral System Series Book 3 – Horse Property Planning and Development* for lots more information about fencing.

—

Management solutions

Feeding confined horses

This book does not cover the subject of feeding horses in detail, but this section gives some pointers to keep in mind for confined horses.

The term 'ad-lib' means that something is provided on an 'all you can eat' basis. In the case of hay provision, it means that a horse always has hay available as opposed to being fed measured amounts. It might sound crazy to feed a horse 'ad-lib', but this is what a horse has evolved to deal with. In the naturally-living situation, they are surrounded by their food and graze in bouts and with periods of rest, rather than eating a 'meal' as a predator does and then having to go without food until they make another kill.

In the domestic situation, we can more closely copy this natural situation of having ad-lib feed by aiming to have low energy ad-lib hay or pasture available for our horses.

Horses that are confined, and therefore unable to graze, must be provided with plenty of fibre to make up for not being able to graze – remember – without fibre, acid builds up in the stomach.

One of the most common (and deadliest) mistakes made by horse owners is to feed their horse as they would feed themselves or their dog - on small but high energy meals. Humans and dogs naturally eat much smaller amounts of higher energy food (relatively). This is because their food types are relatively higher in energy (meat and relatively easy to digest vegetables etc.). Horses are completely different; their food (pasture plants) is difficult and time consuming to digest and therefore, confined horses should be provided with enough hay to allow them to 'graze' as and when they want. As already mentioned, ideally, hay should be provided on an 'ad-lib' (an 'all-you-can-eat') basis when they are not grazing.

Another common horse management mistake is to 'lock horses up' without food in an attempt to reduce their feed intake; this practice is commonly done with horses that are getting fat on pasture. Remember - this is not good horse management because it leads to gorging when the horse is allowed to eat again.

Another common horse management mistake is to 'lock horses up' without food in an attempt to reduce their feed intake; this practice is commonly done with horses that are getting fat on pasture.

If horses have long periods without food, the risks of colic and gastrointestinal ulcers increase and even laminitis can be brought on by the stress caused by incorrect feeding (including 'starving').

Clean but low-energy grass hay is better for feeding horses 'ad-lib'; rather than Lucerne/alfalfa hay, because it is less nutritionally dense. Therefore, more of it can be eaten, thus satisfying the horse's high frequency chewing rates and the guts need to be constantly processing fibre.

If a horse tends to get fat easily, aim to reduce the *energy* value in the hay in order to maintain the quantity hay; for these animals, aim to source hay with a low sugar value. This can be hard to determine, but if you are buying it from a produce/feed store, you need to ask if they have hay that has had a basic nutritional analysis carried out on it (some produce/feed stores will now provide this service). Soaking suspect hay in water (it can then be fed wet) for at least an hour before feeding will help to leach out some of the sugar content.

Be aware that the results of soaking are variable depending on how much sugar there was to start with and the temperature of the water (warm/hot water will leach more sugar). In addition, increase your horse's exercise – this is a very important but often ignored point. Remember - horses are meant to move a lot. It is common for people to go to great lengths to reduce the chance that their horse will develop or have a reoccurrence of laminitis – by 'micro managing' the horse diet. It is particularly surprising that many horse owners will opt to buy expensive supplements and feeds in preference to planning a more naturally active lifestyle for their animals. Increased exercise is a cheaper, more effective way to prevent obesity and it's diseases, it also leads to a mentally more balanced horse. (see the section *Ideas for extra exercise*).

Many people underestimate how much fibre a horse actually needs. An average hay bale (small square) has 10 biscuits (sections) of hay. If a horse is confined for all or most of each day, a medium size (14-15hh) horse needs *at least* 1/3 (3-4 biscuits) of a (heavy compacted) bale to go through its gut daily. A larger horse needs as much as 1/2 of a bale (5 biscuits) or even 3/4 of a bale (7-8 biscuits) of hay per day. This is just a very rough guide, as bales of hay vary very much in weight.

Another rough calculation is that a mature horse needs to eat approximately 2% of its bodyweight in Dry Matter (DM) per day. So a 500kg (1100lb) horse will need 10kg (22lb) of hay (hay does not have much water content so if you are feeding haylage or silage, which does contain water, this figure would be higher). In addition, a horse may need minerals adding to their diet.

These amounts are just to give an inexperienced horse owner a rough idea of the volume a horse actually needs. In reality, a horse should always have access to ad-lib hay when not grazing.

A horse that is working hard may also need supplementary hard feed (e.g. grains or mixes), but be careful as another common mistake that horse owners make is that they tend to overestimate their horse's hard feed requirements.

Remember - horses that are 'group housed' should be able to get out of each other's way and should be separated for supplementary feeding if communal feeding initiates aggression. Horses should ideally be separated into individual yards or stables for the short time that it takes to eat any concentrate feed; both for their own safety and the safety of their handlers.

Changing a horse to 'ad-lib' feeding

If your horse has always had measured/restricted amounts of food rather than ad-lib food then you will have to be careful about changing them over. When horses have been withheld from food they tend to 'gorge' when first allowed to eat at will. Remember - a horse would naturally spend most of its day eating fibre, its whole physiology has evolved to allow it to do this efficiently. When you use restrictive feeding/grazing practices, this is in complete contrast to what the horse has adapted to do and when combined with the horses natural instinct to try to gain weight whenever possible, it is easy to see why many horses develop 'eating disorders'.

So, if the horse is currently living on short stressed grasses (and is overweight), it would *not* be a good idea to switch to turning him or her out on long grasses straightaway; even though these grasses are lower in sugar per mouthful, because the horse in question will initially gorge themselves. **A better strategy would be either of the following:**

Option 1: Over winter – with no access to pasture initially:

- You will need the use of surfaced holding yards – preferably as part of an **Equicentral System**.

- During winter, when the horse is receiving no pasture, feed ad-lib low-energy hay in a surfaced yard – preferably with other horses. You may want to soak this hay in warm water for at least one hour before feeding as a further precaution, particularly if you are not sure what the energy level of that hay is. **An extremely important factor is that the hay does not run out – at all, ever!** This is because, if it does, the horse thinks he or she is being 'starved' again and behaves accordingly (e.g. starts to gorge when food is available again).

- Aim to reduce the horse's weight gradually but significantly over the winter by totally avoiding high energy supplementary feeds, avoiding rugging unless absolutely necessary (but ensure that the horse can get under a shelter) and *increasing exercise* (see the section *Ideas for extra exercise*). You need to aim for a condition score of no more than 2.5 by the start of spring.

Condition scoring

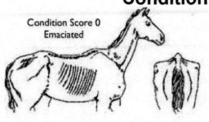

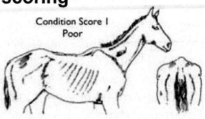

Score 0 (Very Poor). *Neck* - marked 'ewe neck' - narrow and slack at base. *Back & ribs* - skin tight over ribs, very prominent backbone. *Pelvis & rump* - very sunken rump, deep cavity under tail, angular pelvis, skin tight.

Score 1 (Poor). *Neck* - 'ewe' shaped, narrow & slack at base. *Back & ribs* - ribs easily visible, skin sunken either side of backbone. *Pelvis & rump* - sunken rump but skin slacker, pelvis and croup highly defined.

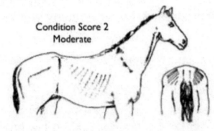

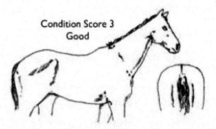

Score 2 (Moderate). *Neck* - narrow but firm. *Back & ribs* - ribs just visible, backbone well covered but can be felt. *Pelvis & rump* - flat rump either side of backbone, croup well defined, some fat, slight cavity under tail.

Score 3 (Good). *Neck* - firm, no crest (except in a stallion). *Back & ribs* - ribs just covered but easily felt, no gutter along back, backbone covered but can be felt. *Pelvis & rump* - covered by fat and rounded, no gutter, pelvis easily felt.

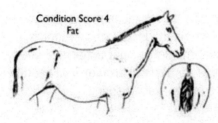

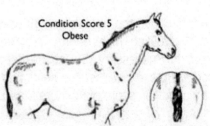

Score 4 (Fat). *Neck* - slight crest. *Back & ribs* - ribs well covered, need firm pressure to feel, gutter along backbone. *Pelvis & rump* - gutter to root of tail, pelvis covered by soft fat - felt only with firm pressure.

Score 5 (Very Fat). *Neck* - marked crest, very wide and firm, lumpy fat. *Back & ribs* - deep gutter along back, back broad and flat, ribs buried cannot be felt. *Pelvis & rump* - deep gutter to root of tail, skin is distended, pelvis buried under fat.

- The horse can still be given minerals etc. if you feel that they are needed, but these do not have to be added to high calorie feed; they can be added to a small amount of chaff.

—

We believe that *not* allowing this naturally occurring weight loss to happen in winter is one of the primary reasons for the obesity epidemic today.

—

- If your land recovers enough to grow pasture before the winter, and this paddock is locked up, you can start to introduce the horse gradually to the pasture over the latter part of the winter. Letting pasture grow long and then allowing horses to harvest it themselves in winter is called 'foggage' or 'standing hay'. This practice has many, many advantages including that it saves the costs of cutting, baling and storing hay (and the risk of it 'failing' as a crop). By mid-winter, it will have more fibre value than nutritional value; in other words, it is ideal for a 'weight challenged' horse.

—

See the second book in this series *The Equicentral System Series Book 2 - Healthy Land, Healthy Pasture, Healthy Horses* for more information about this practice.

—

- When the rest of your land is ready to receive horses again in the spring, you can gradually allow the horse in question to have at first one grazing turnout session (grazing bout) per day over a period of about a week (in addition to ad-lib hay), then allow this session to be longer (for about a week) and so on.

- Extra exercise may be necessary during this period too and it is extremely beneficial if you can do this (see the section *Ideas for extra exercise*).

- By using **The Equicentral System** you will be able to, at first, dictate when that first grazing period takes place; very late evenings (well after sundown) or very early mornings are a good time – but you will need to bring them back in by lunch time at the latest. Initially, avoid allowing the horse to graze between mid-day and nightfall, because this is when the sugars in the grasses are at their highest levels.

- By late spring/early summer, as long as you are keeping a good watch on the horse's body condition score and are not reverting to restricting their low energy hay intake, you should be able to allow night and day grazing bouts with free access to the paddock that is currently in use.

- By the time the pasture is growing higher energy feed in the spring, the horse will have relaxed and will not be as tempted to gorge. If you have carried out the above steps, the horse will now have a much lower body condition score and will be in a much safer position. You will need to keep up this pattern of reducing the horse's weight every winter and keeping up the extra movement whenever it is necessary.

An extremely important factor is that the hay does not run out – at all, ever!

Option 2 – During summer – with access to pasture.

- This option is if you would like to start changing a horse over to ad-lib feeding right away, without waiting for winter. Again you will need the use of surfaced holding yards – preferably as part of an **Equicentral System**.

- Initially, confine the horse by day, on ad-lib low energy hay. Again, it is imperative that the hay **never** runs out and again, you may want to soak this hay in water before feeding, only allowing one grazing turnout period per day as per the previous example.

- You will need to closely monitor the horse's weight and you should **definitely** increase exercise during this period, which should be easier for you at this time of year (see the section *Ideas for extra exercise*).

- As in the above option, avoid supplementary feeding and rugging. Make sure the horse has access to shade/shelter and they should preferably be kept with other horses. Carry on adding grazing time as per option 1, but **only** if you feel the horse is not increasing weight too fast.

- The idea is that you are initially controlling the horse's intake by allowing ad-lib access to low energy hay, but you are switching the horse over to not feeling restricted *at all*. Remember - restricted feeding can actually increase insulin resistance levels because the body reacts by going into 'starvation mode' - **never lock a horse up without something to eat.**

- Never limit hay, limit grazing time initially if you feel the horse is gaining weight too fast. When winter arrives and for every winter from now on, still aim for the horse to lose some weight, because this is what the horse has done naturally for eons; lost weight in winter and not been in as dangerous a position when the feed quality increases in the spring. They can then relatively safely gain some weight gradually during spring and summer. This is a better strategy than trying to maintain the horse's weight at exactly the same level all year round.

- You may need to learn more about pasture plants too, including factors that make them safer, or not as safe, to graze.

Make sure the horse has access to shade/shelter and they should preferably be kept with other horses.

Source - Gina Sykes of Bathurst Equine Agistment - Aus.

Before embarking on any radical changes such as those outlined above, have the horse checked out by an experienced *equine* veterinarian, preferably someone who has a particular interest in the subject of equine obesity. You could also engage an equine nutritionist; preferably an *independent* equine nutritionist.

—

See the second book in this series *The Equicentral System Series Book 2 - Healthy Land, Healthy Pasture, Healthy Horses* for more information.

—

Ideas for extra exercise

Curiously, most pet owners recognise that if they own a dog they should 'walk' it (even if they do not actually do it) but many horse owners do not apply the same ethos to horsekeeping. Maybe because dogs live in or around a home and tend to remind their owners that they want to go out? Horse owners tend to assume that their horse/s get enough exercise if they are turned out on a pasture but as we now know they usually need more exercise than that, plus they are usually receiving more energy (from pasture/feed) than they are using.

There are various ways, apart from general riding, that you can incorporate *extra* exercise into a horse's management routine.

- Driving - small ponies in particular can be trained to drive. Driving is a good pastime for horses and people.

- Hand walking – there is no reason why you cannot take a horse for a walk, smaller ponies in particular are not difficult to do this with. You could even walk the dog at the same time!

- Running – some owners jog/run with their horse in hand for mutual fitness benefits.

- Hill climbing in hand – if you live in a hilly area you can walk/jog/run the hills and you can use your horse to help you up the hills by holding onto their mane.

- Lunging – just ten minutes a day of trotting on the lunge is great exercise.

- Round penning – like lunging but loose in a round yard.

- 'Riding and leading' – if you are riding anyway, why not lead another horse while doing so?

You (and your horse) may need instruction before carrying out some of these activities, but the benefits will be worth it.

Introducing horses to herd living

Careful introduction of a horse into a herd will vastly reduce any risks and an added bonus is that grazing management is much easier when horses live together, because they can be rotated around areas as a group. One horse per paddock does not allow the pasture any rest and recuperation time; a welfare issue for grass! In fact, 'managing' pasture in this way leads to stressed grass that is not good for horses and land degradation problems.

—

We cover this subject in detail in the second book in this series *The Equicentral System Series Book 2 - Healthy Land, Healthy Pasture, Healthy Horses*.

—

If you decide to integrate your horses into a herd because of the horse welfare *and* land management benefits, there are several things to think about and steps to take so that the integration goes smoothly. First of all, think about each of the horses in question and decide if it will be best to have one herd or more than one herd. To end up with just one herd is the best outcome because this will be much easier for land management, but this may not be possible in your situation.

Some of the factors that will help you to decide include the age and sex of the individual animals; for example, young and boisterous horses may be too energetic for a *very* old horse whereas, older horses, on the other hand are usually very good at holding their own with younger horses up until a certain age (which is different for all horses), when they may start to need more specialised care in general.

It generally works best if there are more mares than geldings in a herd, because some geldings still have some entire (stallion) behaviour and can become protective of mares to the point that they will chase other geldings if mares are present. So, if you decide to have two herds for example, it may be better in this case to have one gelding with the mares in one herd and the rest of the geldings in another herd. This is similar to the natural groups that occur in the naturally-living situation e.g. a stallion with some mares and a bachelor group consisting of males of all ages. All horses are different though and various scenarios can work.

There are many ways that a new horse can be introduced to a group of horses. Remember that the existing group will have a social structure and the introduction of a new member will temporarily disrupt this. It is simply not safe to turn the new member out into the group and 'let them get on with it'. In a confined space, the new horse can be run into or over a fence by the other horses. It is better to let the new horse get to know at least one member of the group in separate, securely fenced yards (preferably 'post and rail') or stables that have an area where two horses can safely interact with each other. This way, the newcomer can approach the fence or wall to greet the other horse, but can also get away if necessary. There will usually be squealing, but this is perfectly normal behaviour when horses meet and greet each other.

It is better to let the new horse get to know at least one member of the group in separate, securely fenced yards (preferably 'post and rail') or stables that have an area where two horses can safely interact with each other.

Once these two horses are accustomed to each other, you can turn them out together and then add other herd members gradually one at a time. Try not to give them any hard feed (if the horses are being supplementary fed), only hay, just before you turn them out so that they get down to grazing sooner. It is safer if the horses are left unshod at least for the initial introductions. Hoof boots can also be used. Make sure that resources are plentiful and easy to reach, for example, situate the water away from a corner so that they can each drink safely and no one gets trapped. The group must be watched very carefully during these times.

Keep in mind that you will usually see the most excitable behaviour between the horses in the first hour or so after turning them out together. It can be challenging when introducing horses to one another for the first time, but as long as it is done in a safe manner, usually after a short, somewhat noisy and lively period things settle down. There is bound to be some initial excitement when two horses meet for the first time, particularly if they have been previously kept alone or are new additions to an established herd. Think about when introducing new cat or dog to an established household, there are normally one of two skirmishes until everyone gets used to each other. This is no different in the horse world, but because they are large, energetic, noisy and valuable animals we are often shocked by the first encounter. We have to be aware that some of that behaviour is horseplay; by its definition in human terms, loud and boisterous, but remember this is what horses do. However, we have to be aware of the difference between play behaviour and aggressive behaviour, and ensure that we monitor the situation. With some horses this can be very difficult, as they have been mentally damaged by the way they have been raised or separated in the past.

The Equicentral System - in conclusion

As we have shown, there are many challenges facing modern horse owners. Traditional systems are not meeting the needs of our modern horses. We have to look at more holistic management systems that can work within the boundaries and limitations of our and that of horse's lifestyles.

The simplest way to achieve this is to accept that although we cannot provide a fully natural lifestyle for our domesticated horses, we can learn from nature and work with it rather than against it. Once we start to do this, everything becomes easier, more productive and healthier. We should also invest time in learning about what our horses actually need, this way we are better equipped to make informed choices not only about how we manage our horses but also in who we turn to for advice when needed.

By becoming more responsible, sustainable and ethical horse owners we can ensure that we strive to create an environment which is conducive to creating healthy land, which then creates healthy pasture to ensure we have healthy horses.

Further reading - A list of our books

Buying a Horse Property

Buying a horse property is probably the most expensive and important purchase you will ever make. Therefore, it is very important that you get it right. There are many factors to consider and there may be compromises that have to be made. This guide to buying a horse property will help you to make many of those very important decisions.

Decisions include factors such as whether to buy developed or undeveloped land? Whether to buy a smaller property nearer the city or a larger property in a rural area? Other factors that you need to think about include the size and layout of the property, the pastures and soil, access to riding areas, the water supply, and any possible future proposals for the area. These subjects and many more are covered in this book.

A useful checklist is also provided so that you can ask the right questions before making this very important decision.

If you are buying a horse property, you cannot afford to miss out on the invaluable information in this book!

The Equicentral System Series Book 1: Horse Ownership Responsible Sustainable Ethical

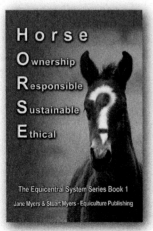

With horse ownership comes great responsibility; we have a responsibility to manage our horses to the best of our ability and to do this sustainably and ethically.

Horse keeping has changed dramatically in the last 30 to 40 years and there are many new challenges facing contemporary horse owners. The modern domestic horse is now much more likely to be kept for leisure purposes than for work and this can have huge implications on the health and well-being of our horses and create heavy demands on our time and resources

We need to rethink how we keep horses today rather than carry on doing things traditionally simpl because that is 'how it has always been done'. We need to look at how we can develop practices that ensure that their needs are met, without compromising their welfare, the environment and our own lifestyle.

This book brings together much of the current research and thinking on responsible, sustainable, ethical horsekeeping so that you can make informed choices when it comes to your own horse management practices. It starts by looking at the way we traditionally keep horses and how this has come about. It then discusses some contemporary issues and offers some solutions in particular a system of horsekeeping that we have developed and call **The Equicentral System.**

For many years now we have been teaching this management system to horse owners in various climates around the world, to great effect. This system has many advantages for the 'lifestyle' of your horse/s, your own lifestyle and for the wider environment - all at the same time, a true win-win situation all round.

The Equicentral System Series Book 2: Healthy Land, Healthy Pasture, Healthy Horses

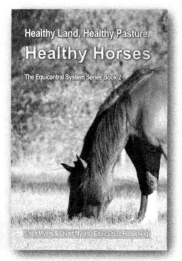

If you watch horses grazing pasture, you would think that they were made for each other. You would in fact be correct; millions of years of evolution have created a symbiotic relationship between equines (and other grazing animals) and grasslands. Our aim as horse owners and as custodians of the land should be to replicate that relationship on our land as closely as possible.

In an ideal world, most horse owners would like to have healthy nutritious pastures on which to graze their horses all year round. Unfortunately, the reality for many horse owners is far from ideal. However, armed with a little knowledge it is usually possible to make a few simple changes in your management system to create an environment which produces healthy, horse friendly pasture, which in turn leads to healthy 'happy' horses.

Correct management of manure, water and vegetation on a horse property is also essential to the well-being of your family, your animals, your property and the wider environment.

This book will help to convince you that good land management is worthwhile on many levels and yields many rewards. You will learn how to manage your land in a way that will save you time and money, keep your horses healthy and content *and* be good for the environment all at the same time. It is one of those rare win-win situations.

The Equicentral System Series Book 3: Horse Property Planning and Development

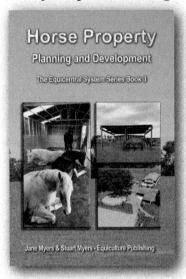

It does not matter if you are buying an established horse property, starting with a blank canvas or modifying a property you already own; a little forward planning can ensure that your dream becomes your property. Design plays a very important role in all our lives. Good design leads to better living and working spaces and it is therefore very important that we look at our property as a whole with a view to creating a design that will work for our chosen lifestyle, our chosen horse pursuit, keep our horses healthy and happy, enhance the environment and to be pleasing to the eye, all at the same time.

Building horse facilities is an expensive operation. Therefore, planning what you are going to have built, or build yourself is an important first step. Time spent in the planning stage will help to save time and money later on.

The correct positioning of fences, laneways, buildings, yards and other horse facilities is essential for the successful operation and management of a horse property and can have great benefits for the environment. If it is well planned, the property will be a safer, more productive, more enjoyable place to work and spend time with horses. At the same time, it will be labour saving and cost effective due to improved efficiency, as well as more aesthetically pleasing, therefore it will be a more valuable piece of real estate. If the property is also a commercial enterprise, then a well-planned property will be a boon to your business. This book will help you make decisions about what you need, and where you need it; it could save you thousands.

Horse Properties - A Management Guide

This book is an overview of how you can successfully manage a horse property - sustainably and efficiently. It also complements our one day workshop - *Healthy Land, Healthy Pasture, Healthy Horses*.

This book offers many practical solutions for common problems that occur when managing a horse property. It also includes the management system that we have designed, called - **The Equicentral System**.

This book is a great introduction to the subject of land management for horsekeepers. It is packed with pictures and explanations that help you to learn, and will make you want to learn even more.

Some of the subjects included in this book are:

The grazing behaviour of horses.

The paddock behaviour of horses.

The dunging behaviour of horses.

Integrating horses into a herd.

Land degradation problems.

The many benefits of pasture plants.

Horses and biodiversity.

Grasses for horses.

Simple solutions for bare soil.

Grazing and pasture management.

Grazing systems.

Condition scoring.

Manure management... and much more!

A Horse is a Horse - of Course

Understanding horse behaviour is a very important part of caring for horses. It is very easy to convince yourself that your horse is content to do all of the things that you enjoy, but a better approach is to understand that your horse sees the world quite differently to you, after all, you are a primate (hunter/gatherer) and your horse is a very large hairy herbivore! So it's not surprising that you both have a very different view of the world.

A good approach is to take everything 'back to basics' and think about what a horse has evolved to be. This book describes horse behaviour in both the wild 'natural' environment and in the domestic environment. It then looks at how you can reduce stress in the domestic horse by understanding and acknowledging their real needs, resulting in a more 'well-adjusted', content and thriving animal.

Do your horse a favour and read this book!

Horse Rider's Mechanic Workbook 1: Your Position

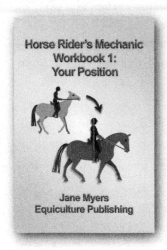

Horse Rider's Mechanic Workbook 1: Your Position

Jane Myers
Equiculture Publishing

Many common horse riding problems, including pain and discomfort when riding, can be attributed to poor rider position. Often riders are not even aware of what is happening to various parts of their body when they are riding. Improving your position is the key to improving your riding. It is of key importance because without addressing the fundamental issues, you cannot obtain an 'independent seat'.

This book looks at each part of your body in great detail, starting with your feet and working upwards through your ankles, knees and hips. It then looks at your torso, arms, hands and head. Each chapter details what each of these parts of your body should be doing and what you can do to fix any problems you have with them. It is a step by step guide which allows you to fix your own position problems.

After reading this book, you will have a greater understanding of what is happening to the various parts of your body when you ride and why. You will then be able to continue to improve your position, your seat and your riding in general. This book also provides instructors, riding coaches and trainers with lots of valuable rider position tips for teaching clients. You cannot afford to miss out on this great opportunity to learn!

Horse Rider's Mechanic Workbook 2: Your Balance

Horse Rider's Mechanic Workbook 2: Your Balance

Jane Myers
Equiculture Publishing

Without good balance, you cannot ride to the best of your ability. After improving your position (the subject of the first book in this series), improving your balance will lead to you becoming a more secure and therefore confident rider. Improving your balance is the key to *further* improving your riding. Most riders need help with this area of their riding life, yet it is not a commonly taught subject.

This book contains several lessons for each of the three paces, walk, trot and canter. It builds on *Horse Rider's Mechanic Workbook 1: Your Position* teaching you how to implement your now improved position and become a safer and more secure rider. The lessons allow you to improve at your own pace, in your own time. They wil compliment any instruction you are currently receiving because they concentrate on issues that are generally not covered by most instructors.

This book also provides instructors, riding coaches and trainers with lots of valuable tips for teaching clients how to improve their balance. You cannot afford to miss out on this great opportunity to learn!

You can read the beginning of each of these books (for free) on the on the Equiculture website www.equiculture.com.au

We also have a website just for Horse Riders Mechanic www.horseridersmechanic.com

All of our books are available in various formats including paperback, as a PDF download and as a Kindle ebook. You can find out more on our websites where we offer fantastic package deals for our books!

Make sure you sign up for our mailing list while you are on our websites so that you find out when they are published. You will also be able to find out about our workshops and clinics while on the websites.

Recommended websites and books

Our websites www.equiculture.com.au and www.horseridersmechanic.com have links to our various Facebook pages and groups. They also contain extensive information about horsekeeping, horse care and welfare, riding and training, including links to other informative websites and books.

Bibliography of scientific papers

Please go to our website www.equiculture.com.au for a list of scientific publications that were used for this book and our other books.

Final thoughts

Thank you for reading this book. We sincerely hope that you have enjoyed it. Please consider leaving a review of this book at the place you bought it from, or contacting us with feedback, stuart@equiculture.com.au, so that others may benefit from your reading experience.